FIXING AMERICA

An Engineer's Solution to our Social, Cultural, and Political Problems

by William M. Taggart IV, P.E.

editors@emerald-books.com

Fixing America: An Engineer's Solution to our Social, Cultural, and Political Problems

LCCN available upon request

Paperback ISBN: 978-1-954779-83-9
Ebook ISBN: 978-1-954779-84-6

List of Tables

List of Figures

ACKNOWLEDGEMENTS

The first acknowledgment must be to my wife, Jessica, who has stood by me through this entire process and been incredibly tolerant and understanding when I have fallen down rabbit holes of data. Her patience in listening to the stories of my discoveries and theories has made this book possible.

To the friends that have reviewed chapters and explained what it is to live their lives to a white, middle-class, Midwest American engineer: my thanks. Their stories have been enlightening and have taught me not only the diversity of America, but the difficulties that our diversity creates.

To the people at Emerald Books, Jessica and Isaac, thank you for taking a word file and a stack of figures and tables and turning it into a beautiful book.

To American industry, which for 35 years employed me and taught me how to sift through mountains of data to find the pattern that would provide the solution—skills that I never thought I would use to write a book about America's problems.

To the engineering mentors, construction supervisors, and process operators who were patient thirty years ago with a young engineer, who at the time had more questions than answers.

To my son, nieces, nephews, and all the children of America: my hope that this book makes the transition through this moment of interesting times in American history both smoother and quicker.

TABLE OF CONTENTS

SECTION I: THE BASIS FOR FIXING AMERICA

SECTION II: CULTURE WAR ISSUES

SECTION III: SOCIAL POLICIES

SECTION IV: ENERGY POLICIES

SECTION V: WELFARE POLICIES

SECTION VI: ECONOMIC POLICIES

SECTION VII: POLITICAL POLICIES

CONCLUSION: THE HOPE FOR MODERATION

APPENDIX

Introduction: Why Write This Book?

America will never be destroyed from the outside. If we falter and lose our freedoms, it will be because we destroyed ourselves.

—Abraham Lincoln

This book began in the shadow of the 2020 presidential election when Americans were just beginning to understand its repercussions. The outgoing president was unwilling to concede, and the incoming president asked for unity and bipartisan support. Some Republican officials had mounted an active campaign to reverse the election, claiming that election fraud had occurred. Normally a transition time, the handover of presidential power had become chaos. A group of Trump supporters stormed the Capitol and disrupted the certification of the election results by Congress. This proved that the divide within American politics had grown into a chasm. There are still large numbers of Republican supporters who believe that the 2020 election was stolen. This isn't supported by any evidence and ignores a number of Republican Secretaries of State who had certified the results of the contested battleground states. This refusal to accept the election results added a massive tension to a country that had already been stressed out by a pandemic. It showed the disfunction that now exists in a two-party system where people on each extreme view the other party with hate and outrage.

We are in a time when people take positions on subjects and defend them as part of their "tribe" considering anyone who disagrees to be the enemy. That is the only way to explain a portion of Republicans believing that an election has been stolen when no evidence has been provided and every court has ruled against President Trump's legal challenges or ignored them completely. But the problem is not limited to the Republicans. Democrats have taken positions on several issues that haven't been fully thought out and could have unintended consequences. One example is the idea of reaching a carbon neutral position by 2050, which hasn't considered what really needs to happen to make that goal, and the resulting costs and changes to society it will require.

The Democrats and Republicans have continued to define policy positions that are farther and farther apart. This leaves most people alone in the middle without a party, forced to choose between two sets of policies they don't completely agree with. In 2020, the Pew Research Center estimated that 34% of American voters identified as Independent. Of the remaining voters, 29% identified as Republican and 33% as Democrat.[1] When polls estimate whether they "lean" Democrat or Republican, this shows that the country is almost split down the middle, however, that doesn't recognize that the middle ground between the two parties has increased in recent years.

An article in the *Atlantic* from May 2014 cited studies that showed that this middle ground was a sizeable block of the voters and were searching for options between the two parties.[2] They are looking to embrace policies that are neither too conservative nor too liberal and looking for a middle ground where politicians compromise more and take extreme ideological views less. In a nutshell, they are looking for a middle-of-the-road policy and politicians who will implement it.

In 2021, even though Democrats controlled both sides of Congress and the White House, they really did not have a mandate or a true majority. That did not bode well for a progressive agenda as there were moderate Democratic senators who could block any legislation. And if a progressive agenda could somehow be rammed through the Congress or enacted through executive orders, a progressive agenda out of step with most of America could risk the election of a Republican president in 2024.

Change 43,000 votes in the 2020 presidential election in the three key states of Georgia, Arizona, and Wisconsin, and you'd have a 269 to 269 tie in the Electoral College, which would have pushed the decision to a vote in Congress. In 2023, change two senators from Democrat to Republican, and Democrat control of the Senate slips away. And while Biden won the popular vote by over 7 million votes in 2020, Trump won the deep south of Texas, Florida, Tennessee, Alabama, Arkansas, Louisiana, Mississippi, and South Carolina by over 4 million votes. The Midwest states of Missouri, Nebraska, Oklahoma, Kansas, North Dakota, South Dakota, Wyoming, Idaho, and Montana went to Trump by over 2 million votes. That lack of a true majority on either side means that what is needed for this country is a middle ground, a moderate set of policies that can be implemented and solve America's problems.

But why did I write this book? Why me personally? Having spent over thirty years as an engineer solving problems for America's industries, I have learned to look for the facts and take what people tell me with some skepticism. Hard data can be altered or misinterpreted, but hard data won't lie to you and try and hide its faults as people will. When I walk into a facility that has suffered an "incident," the people who operate it are shaken, afraid of losing their jobs or, worse, finding themselves in court. Few will tell you about their mistakes or errors in those moments; it's the data and facts you need to seek out and understand. Finding solutions is about looking for patterns in the data while still keeping an open mind and respecting the data, not trying to twist it to support a theory.

That kind of factual, data-driven approach is especially needed right now when politicians and their allies will lie or try to mold the truth to reinforce their position and please their base. We need to go back to the original data and start from there to really understand what is going on.

We also need solutions to be simple. Rube Goldberg contraptions with hundreds of components leading one into another work in videos, but only after lots of failures and repeated efforts to get it right. Policy needs to be simple, clear, and flexible enough to deal with unforeseen changes.

For years, I was the engineer who was brought in to analyze and solve problems in industrial locations around the world. I have spent nights in the rural parts of America, waking up to go to an oil field or an industrial site in the middle of nowhere. Work has also taken me to Europe, China, South Korea, Malaysia, and the highlands of Indonesian Guinea. Having traveled for work and pleasure to many places in the world, I have come to a simple conclusion. I realized that most people are just trying to feed their families and make ends meet.

This epiphany occurred to me while I was standing on a mountainside in Indonesian Guinea, speaking with workers who happened to be Muslim. This was at the same time Muslim protests in the Middle East and other areas were dominating the news in the mid-1990s. I realized that the media covers the radical protesters and violent events. It is quite boring watching Muslim workers and an American engineer wiring up an electrical panel in the mountains. Or watching American workers build an oil field in south Texas with the same engineer. While news networks and social media are obsessed with extremists issuing rage-filled diatribes, most Americans just want a fair country and the ability to feed and clothe their families. In short, they want the American Dream.

In October 2020, I walked away from a job with a mid-sized oil company. I easily could have stayed, put my head down, and kept designing and building the infrastructure that the company needed. But there were questions, simple questions that I felt we weren't getting the answers to but being asked to accept on faith. Is climate change real? What is really going on with immigration? Can we end fossil fuel usage and if so, how fast? What is needed for America? And most importantly how did we get into such a mess with America so divided? And so began my journey through mounds of data and history to find out what was really going on with the issues that were dividing us. And that journey brought me face-to-face with views that were based on cultural myths that have been handed down and reinforced by political parties, that have proven to be suspect. The core focus of this book has been to let the history and data take the book where it needed to go and in some cases that has led me to question long-held beliefs.

The goal of this book is to define solutions to try and address America's problems. A plan that charts a path to a better America and provides more people the opportunity to reach that American Dream.

William M. Taggart IV
Professional Engineer

A NOTE ABOUT THE FACTS AND DATA USED IN THIS BOOK

When data could be easily found and verified from Wikipedia and other public sources, an endnote is not provided. In some chapters, a reference to an agency of the United States government that collects and provides data on the subject of the chapter will be identified, specific references for data will not be provided in the endnotes if they are sourced from that agency. When the data or quote is from any other source and is not cited in the text, an endnote is provided. I welcome any effort to check the data and facts used, but please understand that sometimes there were slight differences from the multiple sources available and I ask that the public accept any differences less than 1%.

Ten percent of the author's proceeds from each book will be donated to Wikipedia to support their continued efforts to remain a non-partisan source of true facts.

All the tables and figures are the author's own work and can be found as Excel spreadsheets on the website fixingamericathebook.com for review.

SECTION I: THE BASIS FOR FIXING AMERICA

The test of a first-rate intelligence is the ability to hold two opposed ideas in mind at the same time and still retain the ability to function.

— F. SCOTT FITZGERALD

At the heart of science is an essential balance between two seemingly contradictory attitudes-an openness to new ideas, no matter how bizarre or counterintuitive they may be, and the most ruthless skeptical scrutiny of all ideas, old and new. This is how deep truths are winnowed from deep nonsense.

— CARL SAGAN

Despite my firm convictions, I have been always a man who tries to face facts, and to accept the reality of life as new experience and new knowledge unfolds it. I have always kept an open mind, which is necessary to the flexibility that must go hand in hand with every form of intelligent search for truth.

— MALCOLM X

1.
Logic, Science, Facts, and Compromise

The words *question* and *quest* are cognates. Only through inquiry can we discover truth.

—Carl Sagan

Policy should be based on facts, science, and logic with an eye toward a final goal while maintaining a balance with America's core principles and values. That's a mouthful, right? But the thing to understand is a policy based on ideology, emotions, or beliefs cannot be defended nor justified to those who don't share the same beliefs. People tend to listen to their own bubble of news and opinions, almost crafted for them. This limits their exposure to other ideas from people outside their bubble. Policy must be based on inquiry, thorough analysis, and a view toward long-term goals, rather than the short-term noise that these media bubbles tend to generate.

This is the biggest problem with the present policy divide: it is based on emotions and not on science or facts. And emotions and beliefs cannot be quantified and debated easily, which leaves little chance to find that common ground. Further, this worsens when the discourse demonizes the opposing party, and those who disagree are portrayed as evil, delusional, or outright insane. This completely stifles discussion, and it is the reason we have had gridlock within Congress and some state legislatures. It also makes compromise impossible. People can come to an agreement with someone who doesn't see things in the same way, but they cannot with someone they view as evil. There's no common ground for people when one side links the other to insanity or delusion. Americans need to learn how to sift through the discussions and recognize propaganda from both the right and the left. They also need to ignore political leaders or speakers who tell them how evil or dangerous the other side is or how that side will destroy

America. No American truly wants to destroy America, but a number of politicians are trying to implement policy based on special interests, beliefs, or stories that don't fully consider the national views or the real benefit or harm for the overall country. Lobbyists have always roamed the halls of government advocating special interests with very little consideration for the overall country. While some point to these lobbyists' abilities to supply special interest money to politicians and political action committees as a new danger, that danger has always existed. The lack of debate, the lack of conversation about issues, and the demonization of those who don't agree with a political belief—these are the true dangers to America.

What is needed are policies based on facts supported by logic that can be rationally explained to all sides. Logic can also show the benefits to all groups. This may involve policies that seem unfair to one group or another, but a clear illustration of how it benefits the overall country (and what compensations or agreements can be made with the slighted group) should be able to encourage a working policy.

To that end, everyone needs to agree that facts supply the basis of understanding. Often, individual emotional stories are brought out to support each side of an argument. As a culture, we are storytellers who dramatize stories to make exaggerated points. How else do you explain three little pigs huddling in a brick house with a wolf at the door? These stories have provided a way of teaching morals and ideas. But currently, stories abound on both sides of every issue. An example is the case of gun control, where examples of homeowners shooting intruders can be matched with stories of mass murders and children getting shot from unsecured guns. Lacking data and analysis, people must decide their position on the issue based on these competing stories and one's personal views and emotions. The facts and corresponding data would identify how often homeowners legally defended themselves versus how many children were injured, how many homeowners had their own gun turned against them, and how many mass murders have been committed. Individual stories are filled with graphic and emotional impact to sway an argument, but the facts and data need to be studied to understand the overall effect and what is truly occurring to the country. Analysis resulting from data needs to shape policy to address the whole country and not the special interests.

There also needs to be consideration that any policy that results in laws must consider how those laws will be implemented and paid for. It is foolish to pass a law that the members of law enforcement can't actually enforce. It might make some politicians and their constituents feel good, but without the ability to enforce it, there is no real benefit. The policy must be thought through, including how it will be paid for, what it will do, how it will be enforced, and the potential unintended consequences. We can't let politicians demand something that leaves others to work out the details, is impossible to implement, results in crippling unintended consequences, or violates the Constitution.

Everyone also needs to recognize that a policy is not going to make everyone happy. Because we have such a diversity of opinions, no policy is going to satisfy everyone. It is important to accept that compromises are necessary. This runs contrary to a strong belief on the far right and far left that compromise means weakness and giving in to any degree to the other side should be avoided at all costs. But how far are both of the extreme sides really willing to go? The fact we have had almost dysfunctional government for the last fifteen years seems to indicate pretty far.

A definition of democracy is the "control of an organization or group by the majority of its members," and we need to recognize that once the majority has decided, then the minority needs to respect the decision. At the same time, the majority cannot violate America's basic values of freedom and equality and understand that policies cannot discriminate against any group by race, sex, age, economic class, or education.

Compromise in addition to policy based on logic, science, and facts is what America was founded on. These attributes have sustained it during many a crisis and are needed again.

2.
The Three Aspects of America

The happiness of society is the end (goal) of government.
—John Adams

Any policy must consider the separate parts of the country. While we are aware of the political separation between Democrats and Republicans, there are other divides, including the three fundamental aspects of the United States of America: the government, the economy, and society. Like a three-legged stool, they each support and create what we call the United States of America. Remove any one and the country doesn't exist; weaken any one and the country is weakened.

The government provides security, common infrastructure, laws, and their enforcement. The economy is comprised of the individual entrepreneurs, companies, and businesses that provide the industrial output of the country, and they also provide jobs, welfare, and material for the citizens. And finally, the overall society is a combination of the morals, education, culture, and general happiness of the United States of America. Each of the three influences the others and can be influenced by the others, but each one does not totally control the others. A government cannot control society or the economy without becoming a totalitarian regime, which invariably doesn't have a good result for either the economy or society.

An example of this is trying to control or restrict cultural change with laws. In the 1950s, several local jurisdictions (Santa Cruz, California; San Antonio, Texas; and Asbury Park, New Jersey, to name a few) tried to outlaw rock and roll music. There were attempts at censorship, and in 1985, the Parents Music Resource Center founded by Tipper Gore, Al Gore's wife, published lists of objectionable songs. There are numerous examples of trying to suppress or block cultural changes, and unless the government is willing to be extremely repressive (think North Korea), these efforts tend to fail. An attempt by the government to stop or alter cultural change is an attempt to restrict society or even turn back the clock. The truth is that society is changing, always changing, and government needs to adapt to society, not the other way around.

And while we speak of the three general aspects, there is enormous diversity within each aspect. The economy has manufacturing, agriculture, service jobs, professionals, and labor unions that give it not only diversity, but also differing priorities and opinions. The same goes for society, where the last election highlighted the differences between the urban and rural areas and the differences between college- and non-college-educated voters.

Priorities in the government are also diverse. From the State Department focusing on foreign affairs to the Internal Revenue Service focused on revenue and the Department of Education, each one demands resources to accomplish its specific task.

There is also complexity in the relationship between the aspects. An example is the relationship between the government and the economy is taxes. Taxes finance the government to do its business, but high taxes can inhibit the economy, thereby reducing the income and creating unhappiness within society. Low taxes may encourage the economy but may not provide enough cash for the government to fully function. So, any policy must consider the impact to all the aspects of America (government, economy, and society), and consider the unintended consequences.

Another thing to consider is changing demographics. Not only has the population increased, but it is living longer and has changed in fundamental ways. In 1950, the population was 151 million; New York was the most populous state; the median age was thirty years old; 63% of the population lived in urban areas; and 10.3% were non-white (note: Latinos were not tracked separately from whites in the 1950s). By 2010, merely sixty years later, the population had doubled to 308 million, California was the most populous state, the median age had risen to over thirty-six years old, 80.7%

of the population lived in urban areas, non-white (as defined in 1950) had grown to 25.8%, and when Latinos are considered, that changed to 36.3%. The 2020 census has raised the total population to 331 million, the median age now stands at 38.5 years old, and the non-white population has grown to 42.7% of the population. Some may want a return to previous times, but the one constant that must be recognized is change. Everything is changing and nothing is constant.

The last thing to consider about the different pieces of America is the inertia within each one. None of the aspects will react well to massive change in a short time. This is fundamentally true of America, from large organizations down to individuals, big changes in a short time produces a destabilizing effect. It didn't take America days to generate the problems that we now face, nor reach the level of dysfunction that exists between the political parties. It will not be resolved in mere days either. Like a giant ocean vessel, any effort to drastically change course needs to be done with a constant and steady effort over multiple presidencies. It will take time and patience to change, and it will take a set of policies that most Americans can agree on, but we first need to identify which problems need to be solved and what direction we need to go towards.

3.
Climate Change

How could I look my grandchildren in the eye and say I knew what was happening to the world and did nothing?

—Sir David Attenborough

With massive wildfires in Western states, hurricanes setting new benchmarks for coastal damage, and high temperatures in the summer that shatter records set only a few years before, more Americans today think that the climate is changing. And while some may talk about "saving the earth," the earth will be just fine. What we are talking about is saving our way of life. Humans, animals, and plants will evolve and adapt to the changing climate, but civilizations may not. There are numerous examples from history of civilizations changing their environment by over-farming or deforestation and weakening themselves to the point of collapse. The people, plants, and animals went on. They survived and adapted to the new climate, but the civilization fell apart because it could not adapt to the changing circumstances. It's not that change can't happen fast, it's that civilizations cannot always survive sudden changes. Societal norms break down, and the civilization fails. So, what we are really looking at is what needs to be done to save the civilization of the United States of America due to climate change. But first: what are the facts and is there a problem?

While scientists have pronounced that worldwide temperatures are rising, that hasn't satisfied everyone. But as a check, can we look at local temperatures and see temperatures rising in our own backyards? Yes, but seasonal changes and the constant change of temperature through the days mean we need to look at a long-term scale to see the change in temperature over the years. Since the beginning of the United States, people have been recording temperatures, and all that data is kept by the National Oceanic and Atmospheric Administration (NOAA). NOAA has the National Centers for Environmental Information, which can be queried, and you can gather the maximum temperature each day for the last 120 years for any of the thousands of weather stations located throughout the country. Some organizations (such as Berkley Earth) have taken that data, and through filtering and analysis, have produced charts for cities and regions showing climate change. There is some controversy around the filtering and regional adjustments that have been made to the NOAA data as it tends to reduce the size of heat waves during the years prior to 1950 and emphasize recent heat waves resulting in a very high overall temperature rise.

What I set out to do was to use the raw data directly from NOAA without any filtering or change for counties located in various places around the country and chosen randomly. Then I would see if there was any indication that temperatures are rising. The raw temperature data is what engineers and scientists would call "noisy." Cold years and hot years alternate randomly, so it is difficult to see a pattern when looking year to year. That is why the average over many years was used. The following table shows the average of the maximum daily temperature between 1901 and 1990 compared with 2011 thru 2020. The span between 1991 and 2010 was omitted to highlight the change between the past (1901 to 1990) and the present (2011 to 2020). The intermediate years should be somewhere in-between if the gradual heating of climate change is real and would lessen the difference from past to present if they were added to either group. The table is sorted based on population density with dense urban areas, greater than 500 people per square mile at the top, and then rural, 500 to 100 people per square mile, and finally ultra-rural, less than one hundred people per square mile.

County, State	Average Daily Maximum		Change in Temp	Density people/sq mile
	1901 to 1990	2011 to 2020		
Urban/Suburban, greater than 500 people per square mile				
Dallas County, Texas	76.36 °F	78.31 °F	1.95	3018.9
Roanoke, Virginia	66.87 °F	68.10 °F	1.23	2333.7
Fayette County, Kentucky	64.83 °F	66.70 °F	1.87	1041.6
Rural, less than 500 people per square mile				
Ada County, Idaho	64.35 °F	64.52 °F	0.17	372.6
Brazoria County, Texas	79.07 °F	80.05 °F	0.98	275.6
Cole County, Missouri	66.67 °F	66.50 °F	-0.17	194.9
Napa County, California	71.83 °F	73.51 °F	1.68	174.6
Fayette County, Pennsylvania	63.01 °F	63.05 °F	0.04	163.6
Parker County, Texas	75.93 °F	77.09 °F	1.16	129.5
Ultra-Rural, less than 100 people per square mile				
Newton County, Missouri	70.35 °F	69.58 °F	-0.77	93.2
Terrebonne Parish, Louisiana	79.88 °F	79.09 °F	-0.79	89.7
Lane County, Oregon	63.35 °F	63.67 °F	0.32	83.9
Guernsey County, Ohio	63.66 °F	64.45 °F	0.79	74.5
Deschutes County, Oregon	60.13 °F	60.96 °F	0.83	64.7
Highland County, Ohio	63.08 °F	64.14 °F	1.06	61.8
Platte County, Nebraska	62.29 °F	61.47 °F	-0.82	48.9
Dickinson County, Mississippi	53.05 °F	53.18 °F	0.13	33.4
Gage County, Nebraska	63.82 °F	64.03 °F	0.21	25.0
Ripley County, Missouri	70.76 °F	69.40 °F	-1.36	21.3

TABLE 1. TEMPERATURE CHANGES BY COUNTY OVER THE TWENTIETH CENTURY

Author's analysis of maximum daily temperature from the NOAA National Centers for Environmental Information databases, accessed via ncdc.noaa.gov

Some may ask why people in very rural areas don't believe reports that climate change is happening. Part of the reason might be that some of them are not seeing it locally. In some ultra-rural areas, climate change is barely noticeable. A few rural and some ultra-rural counties show cooler temperatures in the last ten years than in the earlier times. Platte County, Nebraska, appears to have cooled, but some of this is driven by high temperatures in the 1930s during the Dust Bowl when droughts and crop failures created an agricultural disaster and affected parts of Oklahoma, Nebraska, and surrounding states. Terrebonne Parish is located on the coast of Louisiana, and lakes and swamps cover a lot of the parish (42% of the parish is considered water). The thought is that all that water may have limited the impact there.

The randomly chosen counties didn't show a clear pattern, and as a result, seven states were checked completely and are listed in the table below.

State	Temperature Change, TMAX Average 1901 to 1990 compared to 2011 to 2020 (Degrees Fahrenheit)					Density of people per Square Mile	GDP per capita (2019 Rank)	Energy per capita (2018 Rank)
	Percentage of Counties Reporting	Urban	Rural	Ultra-Rural	Population Weighted Average			
California	75.9%	2.52	1.40	1.29	2.13 °F	253.5	5	48
Massa-chusetts	71.4%	1.50	1.05	N/A	1.47 °F	883.3	2	44
Texas	15.7%	1.65	1.16	0.75	1.40 °F	107.5	19	6
Ohio	36.4%	0.97	0.20	0.65	0.68 °F	286.0	27	23
Louisiana	25.0%	0.72	-0.18	-0.28	0.08 °F	108.1	33	2
Missouri	32.2%	0.15	0.50	-0.62	-0.01 °F	89.2	37	27
Arkansas	21.3%	1.23	0.00	-1.23	-0.43 °F	58.0	49	17

TABLE 2. TEMPERATURE CHANGES FOR SEVEN STATES

The "Percentage of Counties Reporting" column represents counties that had over seventy-seven years of data from 1901 to 1990, which was the threshold for a county to have a large enough data set to use for comparison. Texas had the fewest at 15.7% (forty of 254 counties) because many counties didn't have weather stations that were operational prior to 1920. California had the most at 75.9% (forty-four of fifty-eight), but several counties lacked complete data sets. The "Population Weighted Average" was calculated by taking the percentage of the population in an area (94.42% of Massachusetts is urban) and multiplying it by the temperature change for each type of area (1.5 °F for urban in Massachusetts) and then adding the urban, rural, and ultra-rural numbers to get the weighted average.

Looking at the results, it is surprising that energy use doesn't correspond to heat rise, but gross domestic product does. The wealthier states of California and Massachusetts show more heat rise than the poorer states of Missouri and Arkansas. Population density also appears to be a factor.

This check of temperature data showed that in most rural and urban areas, there is a significant rise in the maximum temperatures when comparing the present (2011 to 2020) to the past (1901 to 1990). Some may argue that the changes are very small, but that 0.98 °F change in rural Brazoria County, Texas, translates into having eleven more days above 90 °F than in the past. In 1901 to 1990, Brazoria averaged seventy-eight days above 90 °F per year; between 2011 and 2020, it averaged eighty-nine days above 90 °F. And not only that, but those days above 90 °F were even hotter. That is more days of air conditioning units running and using more electricity than in the past.

Some have argued that changes in the heat from the sun or something else could have driven this. But for the last 120 years, solar radiation has remained constant, so that can't explain the change in temperature. But fossil fuel consumption has changed. In 1950, US oil consumption was ~6.5 million barrels per day; in 2019, it was over 16 million barrels per day.

The Energy Information Administration (eia.gov) of the Department of Energy tracks the data on America's energy production and consumption and the facts on fossil fuel consumption are:

- Oil consumption is measured in barrels of oil. A barrel is forty-two gallons. US oil refineries consume about 16 million barrels per day of liquid crude oil to produce petroleum products, such as gasoline and diesel. Each barrel when burned generates ~5.8 million British Thermal Units (BTU) of heat.
- Natural gas is measured by volume of a cubic foot of gas at a standard temperature of 60 °F and atmospheric pressure at sea level (14.73 psi absolute), which is referred to as a standard cubic foot (SCF). Natural gas consumption varies during the year based on seasonal temperatures (more in the winter, less in the summer), but the United States averages ~100 billion SCF per

day. Each cubic foot of natural gas is about 1037 BTUs of heat when burnt.

- Coal is measured in tons (also called short tons), which is 2,000 pounds of weight as opposed to the metric *tonne* or long ton, which is about 2,200 pounds. Coal consumption is about 1.6 million short tons per day in the United States and a short ton of coal when burned generates 19.3 million BTUs of heat.

A BTU is used to measure the effect of heating or cooling and is defined as the amount of heat needed to raise one pound of water by 1 °F. Air conditioning and heating systems are usually sized in BTUs, and for an average house, ten to thirty BTUs per square foot is required to cool or heat it, which in Southern states is typically a 30,000 to 35,000 BTU unit for a 2,000-square-foot house.

Whether through direct heating or conversion to electricity, the bulk of fossil fuel energy ends up as heat. An electric power plant may generate electricity with 30–50% of the hydrocarbon energy it takes in, but the rest ends up as heat out the smokestacks. Once the electricity is generated, 5% is lost through the power lines during transmission to consumers.[3] The lost electricity ends up as heat through the resistance of the power lines to the electric current passing through them. And when the electricity reaches consumers, it generates heat through the electronics, the light emitted, and every other use. As a result, very little of the hydrocarbon energy doesn't end as heat, but all of it ends up as additional carbon dioxide and other pollutants in the atmosphere.

To see this in action, while your phone is charging, feel the phone charger sticking out of the electric socket in the wall, preferably without touching the prongs of the plug and shocking yourself. The wall socket has alternating current (AC), while the phone requires direct current (DC) to charge. The charger converts the power running through the electrical cables in the wall to the socket from AC to DC and in the process wastes some electricity while converting, resulting in the heat you feel emanating from the phone charger. Wired phone chargers send about 80% of the electricity into the phone to charge its battery, and the remaining 20% ends up as heat released into the atmosphere. Wireless phone chargers are even worse with almost half the electricity wasted and converted to heat before it ever reaches the phone. This is true of all electronics; they generate heat to do the services that they provide.

This is also true with transportation. Gasoline engines are 20–35% efficient, because 65–80% of the energy is emitted as heat from the car's tailpipe and radiator. Even the energy going into moving the car ends as heat from the friction of the tires on the road, the friction of air passing over the car, and finally the friction of brake pads, which convert momentum into heat to stop the car.

All the energy we use tends to end up as heat in the environment. So, every day in 2020, the United States generated heat from the following hydrocarbon sources:

Oil	92,800,000,000,000 BTUs
Gas	103,700,000,000,000 BTUs
Coal	30,880,000,000,000 BTUs
Total	**227,380,000,000,000 BTUs**

The area of the United States is 3,797,000 square miles, which means we generate about 60 million BTUs of heat for every square mile every day. And that is averaged across the entire country. The reality is that people and the heat generated are concentrated in urban areas where the amount of heat can be over 2 billion BTUs per square mile every day.

Seems like a lot, but it is a fraction of the heat that the sun supplies every day. Scientists use 1,361 watts per square meter (W/m^2) as an average for the equator, and for a normal day in the United States, that works out to ~4,000 watt-hours/m^2 per day. That 60 million BTUs per square mile of wasted heat works out to 6.8 watt-hours/m^2 across a 24-hour day, a 0.17% increase in heat over what the sun is generating. Which is why most scientists ignore waste heat as having little impact on climate. But when you look at the urban areas, waste heat can get up to 2 billion BTUs/$mile^2$, that converts into 230 watt-hours/m^2, a 6% increase, which may seem small, but can be significant, considering that this is part of a heat balance. For any object, the heat *in* has to balance the heat *out*, otherwise the object gets hotter until they do balance. Increase the heat in, the object temperature increases until the amount of heat it is radiating out matches the amount of heat being applied to the object.

Scientists have been focusing on carbon-dioxide generation, which has risen in recent years with increased human activity around burning hydrocarbons. But as an engineer, the heat produced by burning fossil fuels catches my attention. Explaining climate change from increased greenhouses gases trapping infrared radiation is hard for some people to understand, but everyone knows how to cook, and everyone understands that if you increase the heat on a frying pan, it becomes hotter. Turn up the heat on a frying pan by 6% and the pan gets hotter until the amount of heat it radiates into the air around it increases by that same 6%. Based on the data, we have been increasing the heat on the earth for the last hundred years through the burning of fossil fuels and is that starting to show up in hotter temperatures worldwide?

Remember the 227 trillion BTUs per day for 2020? In 1950, the combined BTU equivalent of fossil fuels was 86 trillion BTUs per day for the United States. We have almost tripled the heat being released in the United States every day in the last seventy years. Not only that, but carbon dioxide levels in the atmosphere were 310 parts per million in 1950, and recent measurements are over 400 parts per million.

The basic fact is that fossil fuels are stored solar heat and stored carbon molecules. They are the result of ancient organic plant material that was created from solar heat and carbon dioxide before being buried deep within the earth. That material was converted due to the pressure of the accumulated rock above pressing down on the organic material and the heat from the earth's core to cook that organic material into fossil fuels. People shouldn't be surprised that releasing all that stored heat and carbon dioxide in fossil fuels might result in rising temperatures.

The small changes in the temperatures listed in the table above may not be enough to convince people who have staked their beliefs that climate change does not exist. But then the question has to be this: if validated data showing an average increase of 1 °F isn't enough, what is? How much damage will it take to convince everyone? And at what point are we beyond the point of recovery? These are the questions that have scientists and climatologists upset, because they don't know where that tipping point is.

Between the years 1400 and 1700, there was what has been dubbed a Little Ice Age where a 1 °F drop in the annual average temperature was recorded in Europe. The records indicate prolonged cold and times of drought documented in crop failures and frozen rivers and other bodies of water that had not frozen in recorded history. Scientists have lots of theories on what caused this drop in temperature, but because this was over 500 years ago, they don't have enough data to be conclusive. The key here is that 1 °F of change had a profound impact on the societies at the time.

So, how much of a rise in global temperatures is needed to produce changes that could impact our society? As a young engineer, a number of my projects involved offshore structures located in the Gulf of Mexico. When I started in the late 1980s, there were engineering standards that defined the maximum wave height that an offshore structure had to withstand. Hurricanes Katrina, Ivan, and many others in the early 2000s redefined those standards, increasing the maximum wave height by over ten feet. In the late 2000s after the destruction of several offshore structures by hurricane-generated waves, oil field operators had to cut the legs off their structures and extend them to increase the height of the decks to protect equipment and personnel from waves. This was not a cheap effort, as it was done offshore working from barges and involved massive hydraulic jacks, cutting the legs of the steel structure, jacking the structure up higher, and then welding leg extensions into the gap of the old legs. The oil companies did this for one reason: it was necessary to preserve those older facilities.

Rolling electrical blackouts in the dog days of summer are another sign of climate change as air conditioning usage is pushing the electrical grid past its limit. Remember those eleven extra days a year of over 90 °F weather in Brazoria County, Texas? Those translate directly into more electricity usage. And that average of eighty-nine days over 90 °F was for 2011 to 2020. In 2022, there were ninety-eight days above 90 °F. The Texas ERCOT electrical market has broken usage records for summer days almost every year.

The conclusion is pretty clear: we are already impacted by climate change with hotter days and stronger storms resulting in increased costs to our society. But there are two arguments being made on why we do not want to change to green energy sources. The first is that in order to protect the economy and keep it strong, we need to continue to use fossil fuels, and the second is that other countries will continue to use fossil fuels even if we stop. The problem with both arguments is that they ignore a basic fact: the United States has only enough proven reserves of oil to last for about ten to twenty years at our present rate of consumption. That means in the future, we will be more dependent on imported oil and vulnerable to foreign countries. If you want American energy independence in twenty years, that is going to depend on having more non-hydrocarbon sources.

Oil is not a renewable resource. Oil is the result of ancient, buried plant material, and it is like a bank account that accrues a very small increase each year, but we are pulling out 13 million barrels per day of oil from the US oil bank account. That account is going to run dry sometime in the future. Not only that, but the worldwide oil account will also run dry at some point, and before that, it will become increasingly scarce. That may seem crazy to people since oil has always been around, but we need to remember that the first oil well was drilled in 1859 in Titusville, Pennsylvania. Every oil field since then has been found, drilled, reached peak production, and then declined into nothing. The famed Spindletop oil field in Texas started production in 1901, produced over 17 million barrels in 1902 (20% of all

US consumption), but by 1905 it had declined to a tenth of its peak production, and by 1936, it was making next to nothing.

So, do you change your energy mix now when you still have some oil in the US bank and can afford to change slowly and make the right decisions? Or do you put if off till later when the change needs to happen rapidly because both the US oil account and worldwide oil account are running out and climate change and energy shortages are impacting civilization? Moving away from oil and natural gas is going to happen at some point, and delaying action only increases our risk to ourselves and to future generations. The increased risk is because sudden dramatic change in never easy and always has unintended consequences.

Any future policy must consider this and the implications for the United States and the world. While the United States has only 4.3% of the world's population, it consumes almost 20% of the world oil supply. Renewable energy (solar, wind, hydroelectric, and biomass) only produces 11% of America's energy needs as of 2019. Hydrocarbons make up 80% of America's energy consumption. Reducing US hydrocarbon consumption is necessary and will be addressed in the section on energy policy. But it must be done in a way that considers the impact on the economy and on society. Any sudden removal of 80% of America's energy will have a major impact on the economy and society. Few of us remember the oil shortages during the 1973 oil embargo, when we were dependent on foreign oil and countries in the Middle East stopped shipping oil to America. A few of us may remember long lines for gasoline after recent hurricanes, but it would be a minor inconvenience compared with the major disruption we are now considering to the national energy supply.

Future policy needs to not only consider the impact of fossil fuels, but it also needs to consider recycling, packaging of products, and waste disposal. Again, with only 4.3% of the world's population, the United States generates 12% of the municipal solid waste, i.e., garbage. While China and India may generate more garbage, their combined populations make up 37% of the world population. On a per person basis, the United States is the largest generator of garbage[4] averaging 4.9 pounds per person per day. That average per person increased until the 1990s and has remained constant since, but that is no reason to celebrate—other countries do not even come close to the pounds per day that Americans generate. The United States has gotten better about recycling and composting, but we still need to address the large amount of garbage generated, which impacts the climate and environment just as much as the rampant use of fossil fuels.

4.
Personal Responsibility

Freedom makes a huge requirement of every human being. With freedom comes responsibility. For the person who is unwilling to grow up, the person who does not want to carry his own weight, this is a frightening prospect.

—Eleanor Roosevelt

One of the principles the United States was founded on is personal liberty. This is the belief that people have the right to make decisions for themselves, but that also means that they have to accept responsibility for their decisions and the consequences that ensue. For a democratic republic, this is a crucial requirement that places authority in the hands of the people to elect the leaders in government and rewards them with the right to conduct their lives with minimal government oversight or intrusion. It should also mean that people who make an effort should enjoy the fruits of their labor. Not that they shouldn't pay more taxes as a result of a higher income, but the system needs to be set up to encourage that extra effort. That extra effort advances society and the economy. To tax people in a way where there is no reward for the extra effort to improve one's life would be both unfair and not beneficial in the long term to the economy or to society.

Personal responsibility was stretched by people refusing to wear masks during the pandemic. This is just one example that brings up the question of what responsibility the individual has to accommodate society's best interests. When do you need to sacrifice your comfort and choice to help society? Where does individual liberty end for society's greater good? The answer is when the expression of personal liberty impacts others. A decision made by one individual with a negative impact on others is actionable not only by society but by the harmed parties who may take the individual to court. So, when the individual's actions harm others, the line is clearly crossed. I think everyone can agree that murder and theft crosses the line and should be punished by the government on behalf of society rather than relying on private groups to sue the individual. Sounds crazy? In England you can mount a private prosecution of an individual on criminal charges, but that has never been the American way.

The problem is that when people balk at the simple request to wear a mask to limit the spread of a pandemic or to get vaccinated, is that enough of a negative action to require a consequence? Very few people object to being required to wear clothes in public, though sometimes local governments try to dictate the types and styles of clothes being worn, which can prove controversial. The objection to wearing masks also appeared during the 1918 pandemic when small groups protested. And in 1918, local governments responded with the same measure of fines, misdemeanor charges, and trying to use police enforcement without much success and with great public uproar. Protestors claimed infringement on their rights and freedom, but masks have been proven to limit the spread of a virus both now and in 1918.

If anyone wanted a roadmap for where the COVID pandemic was going to go, they needed to look no further than the history of the 1918 pandemic. With World War I raging in the background, an influenza virus broke out in the United States and quickly travelled to Europe and then around the world killing millions. It was named the Spanish Flu, because US and European newspapers were restricted by governments at war from covering and reporting bad news. Spain was neutral in the war and their reporters were the first to openly report about the outbreak, hence the name Spanish Flu.

The modern problem is that while the COVID vaccines have shown to be very effective at protecting people from the virus, this hasn't stopped people claiming that they have a right to refuse to take it. The virus has proven fatal to less than 2% of the unvaccinated who come down with it, so there is a potential negative impact to others for not wearing masks or getting vaccinated. But is a 2% chance of fatality enough to justify a criminal consequence?

Forcing people to act in a responsible manner could easily cross over into a limitation of personal liberty. This is seen with any attempt to regulate what clothes people wear or how long their hair can be, which clashes with personal freedom. While we all agree that laws against theft and murder should include punishment, extending those same punishments for personal choices stumbles into a gray area. The issue is that with a clear negative impact, everyone agrees on punishment, but when the impact is unclear or only possible, then punishment becomes difficult, and these punishments can be seen as an infringement on personal choice. The solution is to go back to the basic principle of letting people accept the personal consequences of their decisions.

Such an alternative for refusing to get vaccinated would be to have those who don't get vaccinated to accept the consequences. While the vaccine is free, catching the coronavirus in some cases can lead to a hospital bill that can easily exceed $12,000. So, there is a financial risk and a financial consequence that falls on the unvaccinated and their health insurers. Vaccination status should be provided to all health insurers, and those who aren't vaccinated could be charged more for their health insurance just as smokers are charged more because smoking increases the likelihood of more expensive medical bills. This may seem cruel or outrageous, but it is a consequence of not believing the medical and scientific recommendations and helping to reduce the pandemic's spread. The health insurer didn't make the decision to be unvaccinated, and they should have the option to raise the unvaccinated person's cost of health insurance because there is a financial risk. This is in alignment with the belief that the individual can act as they wish but has to accept the consequences of their action. Because our laws at present don't really address the refusal to get vaccinated, individuals can choose not to be vaccinated, but as they are adding to the risk of medical bills, then they have to accept the consequence of their health insurers charging more.

This leads to another issue: if people don't believe in climate change, how can the government regulate and punish those who won't recycle or reduce their fossil fuel consumption? The solution is the same. If someone wants to own an old 1980s muscle car, that shouldn't be illegal. If someone doesn't want to recycle, that should not be illegal either. But there needs to be consequences, which will be addressed in later chapters.

People need to accept that participating in the civilization called the United States of America means accepting certain responsibilities, such as obeying the laws and paying taxes in exchange for the government protecting their rights and liberties. But that also means that paying taxes makes sense only if individuals are getting value—common defense, infrastructure and the supporting structure of the civilization, courts, administration, etc.

The fact is that policy needs to respect personal liberty, but not interfere with personal responsibility, nor deflect that the individual accepts personal consequences from their decisions. And that society and government need to agree in writing on the rights and responsibilities granted to governments. The present document for that is the Constitution and the laws. And we should take a hard look at them in the wake of the pandemic and the flood of propaganda on social media, things our Founding Fathers could not have foreseen.

5.
America's Safety Net

There is an inverse relationship between reliance on the state and self-reliance.

—William F. Buckley, Jr.

How much of a safety net needs to exist to protect people? What is the responsibility of government and society to provide for those who are disabled or unemployed and cannot provide for themselves? You hear people talking about the government's responsibility for a social safety net, but this is determined by society's commitment to prevent people from falling into poverty. The fundamental problem is that society's commitment reduces personal responsibility. People don't have to be prepared for the potential of unemployment if they know that society will step in. It allows them to have less savings and do less planning for misfortune. This also applies to storms and wildfires where federal disaster assistance has reduced the requirement for people to prepare for disasters.

The best example of this is the National Flood Insurance Program (NFIP) administered by the Federal Emergency Management Agency (FEMA). The NFIP was established in 1968 when private flood insurance became prohibitively expensive. Congress enacted the NFIP to provide flood insurance, and until 2004, it was fully covered by premiums paid by insurance policies to homeowners. But between 2004 and 2012, the NFIP accumulated over $17 billion in debt due to major hurricanes. A law enacted in 2012 aimed at adjusting insurance premiums to reflect actual risks to bring down the debt, but instead in 2014, another law was passed to keep premiums low and prevent the increase of costs to homeowners. In October 2017, Congress cancelled $16 billion of the NFIP debt by absorbing it into the national debt.

This is the government providing for people who live in flood-prone areas and has to be considered as a safety net. There are multiple examples of NFIP and FEMA paying to rebuild homes repeatedly and the question has to be asked: why? Why is society paying to de-risk individuals living in flood-prone areas, especially in view of climate change? As long as the NFIP is subsidized by the government with unrealistically low premiums, people are not going to change their behaviors, and they really are not taking responsibility for where they live. This runs contrary to the American belief in personal responsibility and turns into taxpayer debt.

In a decision in 2021, FEMA rolled out Risk Rating 2.0, where NFIP insurance rates were increased to reflect the actual risk. FEMA estimates that 77% of policyholders will see an increase in flood insurance. Most, 66% of policyholders, will see an increase of less than $10 per month, but the remaining 11% will see higher costs to reflect their actual risk. Why are they doing this? Because by January 2023, the NFIP debt had reached $20.5 billion, even after $16 billion was cancelled in 2017. Also, for 2022, NFIP paid $280 million on the interest for the debt. In 2022, FEMA on behalf of NFIP made multiple proposals to Congress on how to address the debt and put NFIP on a better financial footing. None have been enacted.[5]

A similar problem is in the areas of Social Security and Medicare/Medicaid. In 2019, the United States collected payroll taxes in the amount of $1.2 trillion but spent $2.2 trillion on Social Security and Medicare/Medicaid. This is one of the ticking time bombs hidden within American society. Social Security and Medicare/Medicaid will be discussed in their own chapters, but we need to realize that the present safety net is pushing America's federal budget to a point of collapse that could be reached in the very near future.

The National Flood Insurance Program, Social Security, Medicare, and Medicaid are recent developments to add safety nets. Social Security grew out of efforts by President Franklin D. Roosevelt during the Great Depression to implement federal programs to help a society being ravaged financially. But the safety nets we consider normal were not in existence prior to the 1930s, and even with their perceived necessity, the Supreme Court almost struck them down. In 1930, President Herbert Hoover stated,

"I do not believe that the power and duty of the general government ought to be extended to the relief of individual suffering. . . . The lesson should be constantly enforced that though the people support the government, the government should not support the people."

Of course, Hoover's view of rugged individualism wasn't shared by many as the economy collapsed in the 1930s. He was voted out of office in 1932 in favor of Roosevelt, who promised to address the Great Depression with economic aid. Some are making the case that these safety nets are government overreach based on a literal reading of the Constitution, which doesn't recognize that society has changed to view these safety nets as required. In their own chapters, we will try and figure out if Social Security, Medicare, and Medicaid can be improved to make them more affordable and effective.

Few would argue that we need to end all safety nets, rather the safety nets are beginning to overburden the government and present a future financial risk. To fix America, we have to start with the basic assumption that we do not want the United States to become a "nanny state" where the government provides for and supervises the people. Communist and socialist governments that embraced this view have failed, and it was not how America was founded nor how it has operated. But at the same time, we should not take President Hoover's approach and need to agree that safety nets need to be provided for the elderly, the disabled, and to limit people's exposure to poverty. It should not prevent the exposure to poverty but limit it and give them the opportunity to recover and work to return to normal lives.

If that seems like some kind of balancing act, it is. A government needs to provide enough of a safety net to stop someone's plunge into poverty due to loss of employment or physical disability, to provide a means for people to get out of poverty, but not prevent the individual from continuing into poverty if they so choose. That may seem harsh, but it must be viewed through the prism of limited resources. Spending money on people who do not want to help themselves has to be viewed against spending money on better education for children. Especially when one expense is only holding off a problem, while the other is investing in the future of our country.

6.
Education

If a nation expects to be ignorant and free in a state of civilization, it expects what never was and never will be.

—Thomas Jefferson

A democratic republic needs education.

Let's rephrase that. A healthy and growing democratic republic needs an educated voting public to participate in the democracy in order for it to succeed. A lack of education creates a voting public that is easy to mislead and convince to vote against their long-term benefit. A better-educated voting public would be far less vulnerable to the political propaganda that has been mounted on social media. The existing advertising used by politicians on both sides is not geared to appeal to an educated public with facts and policy, but it is an emotional appeal. Think that isn't the case right now? Spend some time watching political commercials and reading the political appeals on social media.

Political commercials are an emotionally laden dramatic effort to convince someone to vote or donate money. Sound bites and a few facts that may be twisted or outright false are sprinkled in, but the bulk of the political ads are emotional appeals. Even though some studies are showing they have less of an impact than some would imagine, they are all light on facts and policy details, but heavy on emotion.

Not only is education the best means to a better voting populace, but it is also one of the most effective ways to end poverty. The Brookings Institute in 2013 came up with three effective things a person can do to avoid or get out of poverty:

- Get at least a high school education.
- Get a full-time job.
- Wait to get married until after age 21, and delay having children until after being married and financially stable.

The study from the Brookings Institute showed that among people who followed these three simple rules, only 2% ended up in poverty and 75% joined the middle class.[6] And while the article geared these rules toward teenagers, it is the government's role to make sure that an education is available to them.

Our Founding Fathers and important political leaders have recognized the necessity of education. Not only for the support of democracy, but also as a means to improve their lives. As such, improving the public primary and secondary education and ensuring that college education is available to all Americans who can achieve college degrees has to be a priority for fixing America.

7.
The National Debt

No pecuniary consideration is more urgent than the regular redemption and discharge of the public debt: on none can delay be more injurious, or an economy of time more valuable.

—George Washington

The national debt must be considered for any policy. The United States of America, as of June 2023, had a national debt of $31.5 trillion and will be adding to it with the spending that is ongoing as of the writing of this book. The Congressional Budget Office (CBO) predicts further deficit spending through the years, which will drive the national debt to unsustainable levels. The CBO report "Federal Debt: A Primer," issued in March 2020, indicated that the "US debt is high by historical standards and projected to continue rising. High and **rising federal debt increases the likelihood of a fiscal crisis** because it erodes investors' confidence in the government's fiscal position and could result in a sharp reduction in their valuation of Treasury securities, which would drive up interest rates on federal debt because investors would demand higher yields to purchase Treasury securities."[7]

The CBO in the same report said:

> If federal debt as a percentage of GDP continues to rise at the pace of CBO's current-law projections, the economy would be affected in two significant ways: Growth in the nation's debt would dampen economic output over time, and higher interest costs would increase payments to foreign debt holders and thus reduce the income of US households by rising amounts.

The debt breaks down into two main buckets. The first is what is called intragovernmental holdings, which is the money that the government has loaned itself. Sound funny? The best example is the Social Security Trust Fund. Social Security ran surpluses for many years in the past and that surplus was put into a fund that invested in treasury bonds. So, the Social Security Agency has about $2.3 trillion in special treasury bonds. Medicare has a portfolio of treasury bonds as does other agencies. In total, $6.76 trillion of the debt is held by other US agencies. That's about 21.3% of the debt.

The rest of the debt is held by the public, mostly US individuals and corporations, but some is held by foreign interests. The portion of the federal debt held by foreign interests has increased from $3.6 trillion in 2009 to $7.3 trillion in 2023 but has decreased from 47% to 30% as a part of the overall public debt. Still, that means that about 25% of interest payments are going to overseas entities.

Interest payments for the national debt were $475 billion, 8% of all government spending per the CBO for the fiscal year 2022. The US treasury includes other costs and indicates that $574.6 billion was required in 2019, almost 13% of all government spending. And that was with very low interest rates. In September 2019, 61% of the outstanding marketable treasury securities held by the public which constituted $9.9 trillion of the national debt were scheduled to mature in the next 4 years. As securities mature, the government needs to either pay them off or refinance them at the current interest rates. As Federal Reserve interest rates have risen to 5.25% as of July 2023, which is almost to the levels in the 1970s (5.6% average), the US interest payments on the national debt will increase and consume more of the federal budget discretionary spending. The estimate for the interest payment on the national debt for the 2023 budget year is $662 billion.

This is why policy cannot be based on thinking that America has unlimited resources and can spend on every special interest without consequence. The national debt has become that consequence. People may be offended when federal funding for arts and charity is reduced or halts, but what choice is provided? Balanced budgets must become the norm. A balanced federal budget has not occurred in the last twenty years.

Deficit spending during a national crisis (2008 Great Recission and 2020 COVID Pandemic) is likely required, but there is no reason to run deficits during good years as in 2018 and 2019 when $1.76 trillion of debts were incurred in those two years combined. The Congress and the president need to shoulder the burden of this and enact fiscal restraint along with increased taxes and dedicate money to begin paying down the enormous burden of the national debt.

Even in the upcoming years with discussion of moving to a greener economy and infrastructure, government spending has to be tempered with the understanding that deficits cannot be permitted, even in the face of climate change.

In the following chapters, policies will be laid out where the federal government provides funds but requires some partial or full return payment from the beneficiaries of the policy. This may seem strange, but realize it is the same basis that investors use when the return from the initial investment is used to help fund future investments resulting in more investments than was possible initially. The idea here is to invest in beneficial areas for the American society with the idea of getting both positive results and some financial return on the investment to help finance future investment without requiring more federal debt. Something that will make dealing with the massive federal debt more likely while allowing the government to provide for and help American society.

8.
The Principles of Fixing America

The most reliable way to predict the future is to create it.
—Abraham Lincoln

Collected from the previous chapters, the following principles govern the policies for the rest of this book:

- The federal government needs to operate a balanced budget with money set aside to pay down the national debt.
- Decreased use of hydrocarbon energy and increased use of renewable energy needs to be encouraged.
- Increased recycling and decreased waste needs to be encouraged.
- Policy needs to respect personal liberty, but not interfere with personal responsibility.
- Policy should be aimed at improving the overall education of American citizens.
- Policy should be aimed at improving the overall quality of life for all American citizens.

The policies are very simple in this book to allow flexibility, but we need to understand that final policy should also be kept as simple as possible. Using incentives or tax waivers only complicates policy and can create a system that someone will figure out how to take advantage of without providing the benefit that the policy sought. This can be taken to the extreme of Campbell's Law, where measures to correct or provide a metric for a system actually makes it worse. An example is when standardized tests become the pass/fail for not only students, but also a judgment on schools and teachers for funding and advancement. This creates the potential that the school system will focus on teaching to pass the standardized test and the overall education of the student will suffer. That is an example where incentives fail.

The policies laid out are starting points. Final actual policy may require some changes to reach the compromises necessary to get them passed through Congress or local legislatures. But we should not allow special interests to block or twist these policies into unrecognizable shapes.

A key factor to helping you understand the following chapters is to approach them with an open mind and realize that they will challenge concepts and beliefs no matter where you are on the political spectrum. They will also ask you to consider that two things can be true even when they seem contradictory. Such is the nature of the problems that America faces.

The United States has spent resources, energy, and money like the party could go on forever and there's no need to worry about tomorrow. For the sake of our children and our grandchildren, it's time for us to be adults, live within our means, and act for the common good.

SECTION II: CULTURE WAR ISSUES

Culture war: (noun) a conflict between groups with different ideals, beliefs, and philosophies.

—OXFORD LANGUAGES

Men build too many walls and not enough bridges.

—JOSEPH FORT NEWTON

9.
Abortion

The two most important days in your life are the day you are born and the day you find out why.

—Mark Twain

Few topics in recent years have been more polarizing or provoked such enormous emotional reactions as abortion. The pro-life side argues that abortion is murder and demands that all fetuses be allowed to be born. The pro-choice side argues that it is a woman's right to decide if she has a child or not, as a child places a critical responsibility on a woman. And the reason why this remains so polarized and unresolved is that both of those statements are correct. The act of abortion does result in the death of a future child *and* giving birth places an emotional, social, and financial responsibility on a mother who may not be ready to accept that responsibility. Yet when you look at both sides, the focus is on what happens prior to birth. The arguments focus on when does a life begin and what right does a woman have to choose what happens to her, and what right does the fetus have. All these arguments are based in morals and emotions, beliefs that both sides violently disagree about. That makes it extremely hard to quantify, to allow discussion, or to have policy based on data and logic. Let alone have any hope of coming to an agreement or compromise.

But when you talk to both sides and ask if they would like to see fewer abortions, then you finally find common ground. But is outlawing abortion the solution? If abortion was outlawed, what happens next? And it is this, the unforeseen consequence of an abortion ban, that could result in major impacts on society.

So, what are the facts?

- There are ~750,000 abortions per year. In 2018 the CDC reported 619,591 abortions excluding California, New Hampshire, and Maryland, which don't report to the CDC[8], and it is estimated that these unreported areas had approximately 130,000 abortions per year.[9]
- 80–90% are single women.
- 60–70% are below the poverty line.
- >50% are already mothers with at least one child.
- More than 40% are younger than 25.

The facts that jump out are that most are single poor women who already have a child, so a pregnancy and the subsequent child is something they can ill afford. So, who is going to raise the child if abortion is outlawed? The single mother who was willing to have an abortion, a relative to the mother, the father who is likely single and poor or may not even be around, or the government? None of these answers seem likely to succeed and there is a striking correlation that is not discussed. The number of live births in the United States for 2019 was 3,745,540[10], adding another 500,000 to 750,000 to the population each year would be a major increase in the US population.

With the Supreme Court's decision to overturn *Roe v. Wade*, the battle over abortion is now being waged in local legislatures, courts, and ultimately at the ballot box. The results of the state referendums in Kansas and Kentucky, where access to abortion was upheld, and some other races indicate that Republicans may be the dog that caught the car, with no idea what to do now. As we will see later, most Americans favor access to abortion within some limits. The states that had trigger laws go into effect or have passed harsh laws that don't even allow exceptions for rape, incest, or if the life of the mother is in danger are about to find out just how big of a car they have caught.

But looking at what had been done previously, there are federal laws that block any federal funds being used for abortion, the Hyde Amendment since 1980 has barred the use of any federal funds to pay for abortion except to save the life of the mother. There is also the Helms Amendment, which since 1973 has blocked any US foreign aid from being used for abortion. Prior to the Dobbs decision, pro-life-affiliated state legislators were pushing to block funding to Planned Parenthood even for the women's healthcare that they supply that is not related to abortion. The laws in a number of states have become so restrictive towards abortion, that medical providers have no idea what to do or if they could even remove a fetus that has died. If there is such opposition to abortion, how did we get to a point where abortion was legal?

The answer lies in the Republican-appointed Rockefeller Commission on Population and the American Future[11] that published its report in 1972 recommending that abortion be legalized as a means to limit population growth in the United States. The commission was formed by the Nixon administration in July 1969 to examine the growth of the American population and its impact on the American future. This was driven by the fact that the US population in 1940 was 132 million, by 1960 it was 179 million and by 1970, it would be 203 million people. The commission concluded that increased population would have negative impacts to all aspects of American society and the environment. It recommended a legalization of abortion and greater access to contraceptives and sex education. The report stressed that America should pursue a quality of life rather than a quantity of citizens. It also recommended limiting legal immigration to 400,000 per year, stopping illegal immigration and approval of the Equal Rights Amendment. President Nixon even though he had formed the commission in 1969, he rejected the report and all the recommendations.

While Nixon rejected the report, most of the recommendations did get implemented indirectly. The Comstock laws were federal acts passed in 1873 to make it illegal to send contraceptives and other items through the US Postal Service. This spawned state laws that went further to restrict access to contraceptives. In 1972 the Supreme court effectively rendered null and void the restrictions on providing access to birth control. In the following years, most states dismantled their versions of the Comstock laws, thereby allowing more access to contraceptives. The Supreme Court ruling of Roe *vs.* Wade in 1973 legalized access to abortion. Sex education was also expanded in the following years.

One thing not implemented was limiting legal US immigration. In the last twenty years, naturalizations (immigrants allowed to become US citizens) has averaged ~700,000 per year even during the Trump administration and about ~1,000,000 green cards per year were issued in 2019 and 2020. Compare that with ~100,000 naturalizations per year and 373,000 Green Cards during the Nixon administration.[12]

Prior to 1970, average US population growth was 1.5% per year, after the changes in the 1970's, population growth has averaged ~1% per year. That is even with increased legal immigration. Between 2010 and 2020 population growth has averaged 0.7% per year. If US population had continued to grow at ~1.5% per year, the US population would be in excess of 410 million people in 2020. The final 2020 US census numbers put the US population at 331 million. About 80 million less than if the 1.5% growth rate from pre-1970s had continued. Outlawing abortion could have a major increase on US population at a time when due to social issues and federal debt the country is already strained.

Also, the question is what will these children grow into? In a 2001 paper by economists Steven Levitt and John Donohue[13], they argued that passage of Roe *vs.* Wade in 1973 tracked with a decrease in American crime statistics starting in 1992 and dropping sharply in 1995. Their findings have been disputed, but there is no ignoring that a drop in crime did occur in the early 1990s, twenty years after the Roe *vs.* Wade decision. Also, states that were early adopters of legalized abortion saw corresponding earlier drops in their crime statistics. It makes a logical sense that unwanted children have a higher likelihood of becoming, if not criminals, then at least a burden on society.

This leads to the other question of the impact of these children on entitlements. The impact of an additional 750,000 births per year on America's population and costs, both financially and environmentally have not been taken into account. And while abortion of a fetus is a horrific thing, to increase the US population would also increase the demand on government services, resources, and increase the need for jobs. In 1970 it was estimated that the cost to raise a child to eighteen was about $60,000[14], in 2020 the cost of raising a child has risen to ~$233,610[15], not including college. Including college in 2020 (prior to COVID-19) was another $22,000 for a public college (in-state) after factoring in financial aid.

And while people will find it appalling to reduce fetuses and the issue of abortion down to numbers and dollars, as was said at the beginning of the book, policy must be based on facts, logic, and an understanding of the consequences. Prohibition was one example where a minority was able to impose a social change on the country. In 1919, the Eighteenth Amendment was ratified: the "manufacture, sale, or transportation of intoxicating liquors within, the importation thereof into, or the exportation thereof from the United States and all the territory subject to the jurisdiction thereof for beverage purposes is hereby prohibited." By 1933, increased crime, illegal alcohol supply, the loss of revenues from alcohol taxes, and other consequences had reached the point where the Twenty-first Amendment was passed overturning the Eighteenth. The Prohibitionists' failed attempt to transform the social norms showed that the government cannot change society without resorting to draconian efforts with consequences for the health of the society.

Similarly, an outright legal ban on abortion would have far-reaching consequences for the long-term economics, quality of life, and social fabric of America. It is not a policy that can be adopted without considering the costs and the long-term impacts. Consider an additional 500,000 births per year at an average cost of ~$200,000 to raise a child. That is a low estimate of $100 billion per year that someone needs to pay to keep these children out of poverty.

This is further complicated by the fact that 13.2% of the existing live births in the United States took place in households in which the income per member was less than $5,000 per year, far below the poverty line. When you raise the income per household member to $10,000 per year, 31.4% of all births are below that threshold. When you look at households where the income per household member is below $20,000 per year, the number of women of childbearing years (15 to 50) is 44.8% of the total population of childbearing women, but they have 59.9% of the births. Almost 60% of American children are born in households where the income is less than $20,000 per person per year. And 25% of all existing births are from unmarried women making less than $20,000 per person per year.[16]

And what can the world tell us about the pros and cons of abortion? Since the 1950s, seventy-nine countries have legalized abortion for economic or social reasons, where an abortion is permitted in the first trimester if the woman indicates she cannot afford another child. This is 37% of countries and is a slightly greater than the countries that perform abortion only based on a woman's request (34% of the countries). Because our focus is on poverty and its impacts, we are centering on the countries where it is permitted for economic or social reasons. When you compare the poverty, education, and wealth among those who have legalized access to abortion versus those countries where abortion is still illegal, the results are startling.

Countries where abortion is illegal are 56% of the world's population (4.5 billion) but have an overall poverty rate of 25.4% and an average Gross Domestic Product (GDP) of $3,075 per person, when the world average is $10,926 per person. Only 89% of these countries report their high school graduation rates, but among those, about 33% of the adult population have high school diplomas or their equivalent. Of the countries banning abortion, only 47% report the percentage of college graduates in their population and it is only 10%.

Compare that with countries where abortion has been legalized. Those countries are 44% of the world population (3.5 billion), with a poverty rate of 11.8% and an average GDP per person of $20,410, twice the world average. Among the countries that allow abortion, 99% report the percentage of their population with high school diplomas, and it's 77%, more than double that of the countries that ban abortion. This pattern continues into the percentage of the population that is college-educated. Among those that allow it, 85% report that the average is 28% of their population is college-educated, almost three times the percentage where abortion is banned.

Abortion	Population (2022)	Percentage of World Population	Percentage of Citizens in Poverty	GDP per Person	Percentage of High School Graduates	Percentage of College Graduates
Allowed	3,502,688,031	44%	11.2%	$20,448.60	76.8%	27.7%
Prohibited	4,457,403,558	56%	25.6%	$3,075.13	32.9%	10.2%

TABLE 3. HOW ABORTION LAWS CORRELATE TO SOCIO-ECONOMIC FACTORS

Some of the countries that have recently legalized abortion—Argentina (2021), Mozambique (2015), Mexico (2021), and Sao Tome (2012)—have some of the highest poverty rates and lowest education rates among the countries where it is allowed, especially Argentina and Mexico. If those two countries begin an economic and education revival during the next twenty years, it would be strong evidence showing that abortion provides a benefit in reality, regardless of the morality arguments.

We can't know whether rich, educated countries allow abortion or if allowing abortion made those countries rich and educated. One interesting fact is that for most of the countries with lower poverty rates than the United States, they have lower rates of abortion. A Guttmacher Institute study in conjunction with the World Health Organization[17] found the starkest divide in Europe, where among countries with legal abortion and low poverty, rates of abortion were 12.3 per 1,000 women for 2008, while the United States was 20.8 per 1,000 during the same time frame. They estimated that in European countries where it was illegal, abortion rates were 43 per 1,000 women, with illegal abortions also being fatal in some cases.

Another point to make is the change in population. In 1960, the population of the countries that prohibit abortion was 1.2 billion. It has increased to 4.5 billion in sixty-two years, a 274% increase. While the countries that today permit abortion had a population of 1.8 billion in 1960 and are now 3.5 billion, a 92% increase. In Africa, only 9% of the population lives where abortion is permitted, and Africa has experienced over 400% growth, from 272 million in 1960 to 1.4 billion people in 2022 with 600 million people added in the last twenty years. In every region of the world, the population of the countries that ban abortion has grown by a minimum of twice that of the countries where abortion is permitted. In South America, it's almost six times as much.

A last issue from the world view is that the illegal immigration issues on the US southern border stem from countries in Central America, South America, and the Caribbean. If you look at the list of the countries sending the most people to the United States, they all ban abortion, are overpopulated, and have high levels of poverty. Only Mexico, which just had a court ruling allowing abortion, breaks the pattern and as the ruling just happened during the writing of this book, we can't expect to see legalized abortion have any impact on the poverty levels in the immediate time frame.

At the beginning of this chapter, we pointed out that pro-life advocates argue that abortion is the death of a potential child and human being, which is true. They embrace the morality of their position, but without a plan to deal with potential overpopulation or poverty risks, can it be pursued? The other thing to consider is that the pro-life advocates are in the minority. A Pew Research Center poll on abortion in 2019 indicated that 61% thought abortion should be legal in some or all cases, while 26% thought it should be illegal in most cases, and 12% thought it should be illegal in all cases.[18] This further breaks down showing that the areas where pro-life outnumbers pro-choice. In religion, only white evangelicals support pro-life (77%) more than pro-choice (20%). Even Catholics have a larger pro-choice group (56%) than pro-life (42%), and every other major religion leans more towards pro-choice. Along the lines of education, the indications are that college graduates are more pro-choice (70%) than pro-life (30%), while people who haven't attended college are much closer, but still are more pro-choice (54%) than pro-life (44%). The result is that the Republicans are the party of pro-life (62%) versus pro-choice (36%), more so than the Democrats, but it isn't even consistent within the party. Conservative Republicans are more pro-life (77%) versus pro-choice (22%). But moderate Republicans are not in alignment as pro-life is only 41%, outnumbered by the pro-choice at 57%.

So, what is the path forward when most Americans favor legalized abortion? The data shows that banning abortion will have damaging impacts due to poverty and possibly result in higher crime rates in twenty years, but legalization is opposed by a determined group of pro-life advocates that Republicans are enacting legislation to support.

If the objective is to reduce the number of abortions, then a solution is to reduce poverty. But you can't reduce poverty if millions of children are being born into poverty.

With the reversal of *Roe v. Wade*, the issue has moved to the states and the ballot box. The best option is for the pro-life groups to not try and outlaw abortion, but to work to get back to a Roe-like set of laws where abortion is legal up to some point in the second trimester, but to provide provisions for adoption. The idea would be to work to identify families willing to adopt, then screen and pre-approve them and present them as an option for women who want an abortion. And by *families*, I don't only mean heterosexual married couples. This should be expanded to include non-heterosexual couples as well as single people, but they should all be qualified to raise a child and have the means as well as demonstrate that they are emotionally ready. This wouldn't mean a psychological test or social criteria, just that they haven't committed a major felony or have a history of serious mental illness. Think that sounds harsh? There are too many children living in poverty and suffering under the mental illness of their caregivers. To increase that number would be irresponsible. Some may feel that this isn't generous or fair and that more people should be eligible to adopt these children, but we cannot continue to place children into the cycle of poverty and mental illness.

The hope is that pro-choice would accept this idea of a legal but limited abortion with an informal suggestion that Planned Parenthood and other abortion providers are willing to work with this adoption initiative without involving government. This minimizes taxpayer costs and hopefully reduces the tension surrounding this issue. It is not the outright ban that pro-life groups would prefer, but it would not only save some children, but serve as a pathway to a better future for children, avoiding poverty and dependence on a government unable to support them.

Hopefully, this would end the continued reduction and blocking of funds to Planned Parenthood and other healthcare providers, which is an ongoing issue in the federal and state legislatures. Several studies show that these efforts have unintended consequences—negative impacts to women's health in the affected regions. While the reports are still preliminary and inconclusive, an alignment between pro-life groups and abortion providers to encourage adoption and reduce abortions would stop with the burdensome regulations.

If we turn the focus from pro-life versus pro-choice to how to improve the quality of life for Americans, it is clear that a ban on abortion does not do that and in fact is likely to increase the burden of poverty, something that this country cannot afford.

10.
Racism

Hating people because of their color is wrong, and it doesn't matter which color does the hating. It's just wrong.

—Muhammad Ali

There are five issues that I view as being the emotional hot buttons in American society. While abortion was tackled first, that does not make it more or less important than racism. In fact, ranking any of these five hot buttons is an act of futility. They all elicit emotion on both sides and have divided us far more than any others.

Many experts and pundits have said that at the present, we are in a time of systemic racism, but what exactly does that mean? We need to look at our history and our attitudes toward race and how they have changed in the last 200 years. And while this mainly focuses on Black/white relations, it extends to how every race deals with and sees the others.

Racism's basis in America must start with slavery, but not with the Civil War, which marked the official end of slavery. The 1619 Project developed by Nikole Hannah-Jones takes it all the way back to 1619 when a ship with twenty African slaves arrived in Virginia and the slaves were purchased by the European colonists. The 1619 Project attempts "to reframe the country's history by placing the consequences of slavery and the contributions of black Americans at the very center of [the United States'] national narrative."[19] This has been a controversial position. Some people push back on what they are calling revisionist history. As for the historical record, some put the first slaves as far back as 1526 when they were brought into Spanish colonies in present-day Carolinas/Georgia coast, but because these colonies failed, most historians agree with 1619. Regardless, the fact is that slavery was firmly established in all the states prior to the American Revolution. The census of 1800 recorded over 36,000 slaves in what we consider Northern states due to their banning slavery prior to the Civil War. Maryland, Kentucky, and Delaware sided with the Union, but had legal slavery prior to and during the Civil War. Missouri also had legal slavery during the Civil War but didn't exist during the 1800 census as a state. In the same 1800 census, Southern states had over 850,000 slaves to power the plantation-style farming. In Southern states, 33% of the population were slaves; for the overall country, 16.7% of the population were slaves. Which is why the Constitution has the Three-fifths Compromise. Article 1, section 2 of the Constitution states:

Representatives... shall be apportioned among the several states... according to their respective numbers, which shall be determined by adding to the whole number of free persons, including those bound to service for a term of years, and excluding Indians not taxed, **three fifths of all other persons**.

Southern states wanted to count their slaves as part of the population to give them more representation in Congress and the electoral college, but not allow them to vote. Northern states and abolitionists pushed back on this, and the result is the Three-fifths Compromise. None of us were in the room when this was hammered out, but we have to accept that while the word *slave* is avoided, this was an agreement that addressed slaves. And in the 1800 elections, that agreement gave Southern states the ability to control 47% of the electoral college with only 39% of the voters. In the early elections, Virginia had twenty-one of the 138 electoral votes and produced four of the first five presidents.

There has been a lot of concern about what the Founding Fathers did, but what we need to understand is that without the Three-fifths Compromise, there would be no United States of America. Some have said that the Constitution was stained with slavery from the very beginning, but that fails to recognize that the compromise was necessary to form the country. What were the alternatives to the Three-fifths Compromise? The Colonies were exhausted from a war with Great Britain, and there was no interest in fighting a war about slavery at that time. The population was almost evenly split between North and South, and the North had yet to develop the industrial strength that allowed it to win in 1865. In 1800, the North had 653,274 white males, while there were 419,857 in the South. That is not enough of an edge in manpower to invade and win.

By 1860, the Northern population had grown to almost 19 million. Of the 13 million living in slave states, almost 4 million lived in the four slave states that would remain with the Union (Missouri, Maryland, Delaware, and Kentucky), reducing the manpower to 9 million. The North had a two to one population advantage, and it gets even stronger when you realize that over 3 million of the Southern population were slaves. Looking at just white males aged 18 to 45, the advantage in 1860 becomes even starker. The Northern states that fought for the Union had 3.75 million white eighteen-to-forty-five-year-old males while the eleven states that fought for the South had a little over a million. It would not be until the 1860s that the North had the manpower advantage to enable it to fight and win a war against slavery. With that said, without the Three-fifths Compromise, it is unclear if we would even have had a single nation or if America would have ended up either as fragmented states or as two or more nations. Two nations split between free and enslaved, or separate squabbling states, would have made the area vulnerable to European meddling and inter-state fighting. This was something that the Founding Fathers clearly feared, based on what was written in the Federalist Papers.

And we should be cautious about applying modern morals and paradigms to revolutionary days. Even Benjamin Franklin, who wrote on freeing the Black slaves and became president of a Philadelphia abolition society in 1787, owned as many as seven slaves in earlier times and was documented to have two slaves with him when he traveled to England in 1756. We don't totally know what changed Franklin from slave owner to abolitionist, but his exposure to a school for Black children in 1759 and his subsequent donations to the school may mark the beginning.

We know that George Washington owned slaves but by the end of the Revolution had become troubled by slavery. This is indicated by comments left in his papers. In 1788, French abolitionist Jacques Brissot petitioned Washington to establish an abolitionist society in Virginia, and Washington responded that although he supported the idea, the time was not right. Whether he was referring to Virginian society or that he depended on slavery to keep his planation of Mount Vernon working is not clear. Upon his death in 1799, his will directed that the slaves be freed upon the death of his wife Martha, and in the will, he provided a trust to feed and clothe them, which lasted until the early 1830s.

Washington was not a rich man. Mount Vernon barely succeeded even with slaves and removing them would not only have doomed Mount Vernon, but the entire planter economy of the Southern states. And while that partially explains why Southerners could not relinquish slavery, it does not forgive the act nor justify the view that Black people were inferior, which was used to justify the existence of slavery. Pseudo-science and quasi-religious beliefs insisting that slavery was required or natural were held up as justification. There was even a Supreme Court ruling in the 1857 case *Dred Scott v. Sandford*, where the court ruled that people of African descent "are not included, and were not intended to be included, under the word 'citizens' in the Constitution, and can therefore claim none of the rights and privileges which that instrument provides for and secures to citizens of the United States." Prior to the Civil War, slavery was sanctioned by the Supreme Court and in laws throughout the country.

No, if destruction be our lot, we must ourselves be its author and finisher. As a nation of free men, we will live forever or die by suicide.

—ABRAHAM LINCOLN

And then came the Civil War. Some have argued that states' rights was a cause of the Civil War, but that sidesteps the reality of what was happening just before the war. As new states were brought into the United States, the question of whether they were allowed to have slaves or whether they were free states became the overwhelming issue. In the 1856 election, James Buchanan, a Pennsylvania Democrat, won all the Southern states and enough "free" states to become president, defeating John Fremont, the candidate of the new Republican party and Millard Filmore of the American Party, or as it was better known, the "know-nothing" party. The Know-Nothings were remnants of the old Whig party who had not joined the Republicans and avoided the issue of slavery but were fiercely anti-immigrant and anti-Catholic. The Republicans in 1856 were not advocating for the abolition of slavery on a national level but that all new states needed to be free. The South had already started to lose their advantage in the electoral college that the Three-fifths Compromise had provided. By 1856, slave states constituted 120 electoral votes to the free states 175; their 47% had declined to 41% of the electoral college. Historians have theorized that a change of only 25,000 votes would have made the Republican Fremont president in 1856.

By 1860, the situation for the South had worsened with the addition of Oregon and Minnesota, both free states. The 41% of the electoral college had declined to 39.6%. Further, the election of 1860 was held before the results of the 1860 US census had been announced. Those census results would only make it worse for the South in the 1864 elections. Also, because of the Kansas-Nebraska Act, the Kansas territory was to decide if it was free or slave by popular vote. This act sparked the Bleeding Kansas event, where armed conflict between pro-slave advocates and abolitionists in Kansas attempted to skew the popular vote. Faced with worsening odds, Southern leaders drew a line in the sand. If the Republican candidate Abraham Lincoln was elected, it meant secession. When someone says the war was about states' rights, they miss the real argument. The issue of states' rights was not about trade or local jurisdiction, it was about the protection of slave property such that Southern owners could take their slaves into free states.

Dred Scott sued his owner after being taken into a free state and was then returned to a slave state. Scott argued that by taking him into a free state, he was free, but the Supreme Court in 1857 by a seven to two decision did not agree. The decision had been greatly anticipated, and Republicans at the time suspected that President Buchanan had influenced Justice Robert Grier, a Northerner, to join the Southern majority to make the decision appear bipartisan and not split along North/South lines. That suspicion has been confirmed by historians with letters from Buchanan to Grier. But if Buchanan had hoped for a decision to stop the rancor tearing the country apart, he instead got a bombshell from Chief Justice Roger Taney. The Taney decision argued not only that Black people could not be citizens, but also that Congress had no right to block slavery and that any slave brought into a free state remained a slave. Republicans viewed this as an effort to expand slavery into existing free states and only made the partisan divide between the North and South even worse.

And while we mark the firing on Fort Sumter as the start of the Civil War, we overlook the detail that Americans had already been fighting over slavery for six years in Kansas. The reason for the Civil War was that the South depended on slavery (32% of white families in the South owned slaves and 30% of the Southern population was slaves), something that the North had come to view as illegal and immoral.

The transubstantiation of Andrew Johnson was complete. He had begun as the champion of the poor laborer, demanding that the land monopoly of the Southern oligarchy be broken up, so as to give access to the soil, South and West, to the free laborer. He had demanded the punishment of those southerners who by slavery and by war, had made such an economic program impossible. Suddenly thrust into the presidency, he had retreated from this attitude. He had not only given up extravagant ideas of punishment, but he dropped his demands for dividing up plantations when he largely realized that Negroes would be beneficiaries. Because he could not conceive of Negroes as men, he refused to advocate universal democracy, of which, in his young manhood, he had been the fiercest advocate, and made strong alliance with those who would restore slavery under another name.

—W. E. B. DuBois

The era after the Civil War called Reconstruction began with the Federal government's attempt to free the slaves, provide them with "forty acres and a mule," repeal racist laws, and prevent the return to power of Confederate leaders. While Abraham Lincoln set these things in motion, his assassination placed Vice President Andrew Johnson in charge. After assuming the presidency, Johnson attempted to unwind many of Lincoln's efforts. Johnson had been selected by Lincoln to be his vice president on a unity ticket in the 1864 election. Johnson was a Democrat but had shown loyalty to the Union. He had been a slave owner and was opposed to allowing Black people to vote. Congress passed civil rights bills and bills establishing the Freedman's Bureau to help former slaves. These were vetoed by Johnson, but Congress mustered up enough support to override the vetoes and pass the laws. The conflict between Johnson and the Republican-dominated Congress resulted in Johnson being impeached, but when the senate voted, he was acquitted by one vote. The new laws outraged Southerners, resulting in violence against Blacks and northern Republicans (called carpetbaggers) who had travelled South to help them. The Ku Klux Klan was formed as a means of organizing and carrying out the violence. Murders, lynching, and voter suppression occurred, but was somewhat held in check by the presence of Federal troops stationed in Southern states. Johnson campaigned to get troops withdrawn from Southern states and the Union generals presiding over the Southern areas recalled to Washington. The Republican Congress battled back on all these fronts.

When Ulysses S. Grant was elected in 1868, he got Congress to pass the Enforcement Acts, laws designed to allow the Justice Department and Federal troops to go after the Ku Klux Klan. The Klan was essentially driven out of existence so thoroughly, that it only reformed in the 1920s. Unfortunately, tensions within the Republican party and efforts by Northern Democrats meant that by 1874 Democrats were able to gain control of the House of Representatives. The Reconstruction era ended in 1877 when a compromise that elected Republican Rutherford B. Hayes to the White

House also removed the last Federal troops from the South and allowed pre-Civil War Southern political leaders to return to power.

Do not think that the North was innocent in this. The effort to give Blacks the vote and the high cost of Reconstruction created splits in the Republican party. Some Republicans worked with Northern and Southern Democrats and that allowed Democrats to gain control in 1874 of the House and end Reconstruction and begin the Jim Crow era. Under Jim Crow, Black people were supposedly free and allowed to vote, but racist laws and attitudes prevented them from voting and kept them in slave-like conditions in many areas.

Isabel Wilkerson in her book *Caste* argues that the system that arose after slavery in America is not racism, but it is an institutionalized system where Blacks are the lowest caste in America. This is similar to India where a caste system originating in ancient times became a dominant force in society and dictated a person's level based on parentage. If you were born to Brahmin parents, you were in the top class, the priestly rulers. If you were born to Dalits (untouchables), you were on the bottom rung scorned by others and truly untouchable. If a Dalit touched or was touched by a Brahmin, the Brahmin had to go through an elaborate purification ritual and the Dalit was usually just beaten. With the Jim Crow laws, anyone with "one drop" of Black blood was considered Black. There were multiple events cited by Wilkerson where if Blacks swam in a public pool in the 1950s and 1960s, the pool was drained and cleaned before whites would use it, or the pool was simply closed and filled in rather than share it.

What a sad era when it is easier to smash an atom than a prejudice.

—ALBERT EINSTEIN

Einstein is credited with these words, but pinning down when he said them has been difficult. I suspect it was sometime in the 1960s when American society was coming to grips with the elements of Jim Crow that still existed in the South. The Freedom Riders and the Bloody Sunday march over the bridge in Selma, Alabama, brought the underlying racism (or caste as argued by Wilkerson) to the forefront of American televisions. While all eyes were on the South, northern areas were not without issue. Martin Luther King, Jr. took the civil rights effort to Chicago in 1966 with the Chicago Open Housing Movement and received "a worse reception in Chicago than in the South." Racist laws may not have existed in northern states, but racist attitudes did.

Some have argued that inter-racial relationships have improved, going so far as to compare them with countries that are still predominately racist in one form or other, India's caste system being an example. But that doesn't make sense. Claiming that your house looks better than your neighbor's is no excuse to allow the house to go unrepaired and fall apart.

Institutionalized or systemic racism is defined as racism that is normal practice within society, as opposed to when there were laws that actually permitted racism. The last Jim Crow law was overturned in 1967, less than sixty years ago. The Thirteenth Amendment outlawing slavery was ratified by thirty states in 1865. But Kentucky did not ratify it until 1976, and Mississippi did not ratify it until 1995. Both Kentucky and Mississippi originally rejected the Thirteenth Amendment in 1865. The Twenty-fourth Amendment outlawing poll taxes was ratified by enough states to take effect in January 1964, but it was only recently ratified by Alabama in 2002 and Texas in 2009. Eight states, including Arkansas, Georgia, Louisiana, Mississippi, and South Carolina, have never ratified it.

There have been major improvements in racism, considering that people were enslaved 160 years ago, and interracial marriage was outlawed in some states less than sixty years ago. But with a little over two generations from when the last true racist laws were on the books, to insist that racism is gone from the United States is naïve. But there has been improvement. A Gallup poll had been asking the same question for sixty years: do you approve or disapprove of interracial marriage? In the 1950s, 95% disapproved, but the last time they reported the question in 2007 it had fallen to 18%. An improvement, but it still shows that racism remained. Even now, vestiges of Jim Crow still show up. In January 2021, a Louisiana cemetery was found to still list in its sales contracts "white only." After the burial of a Black sheriff's deputy was denied, the cemetery quickly revised its documents.

And yet, in 2019, the Pew Research Center posed the question, "When it comes to racial discrimination, the bigger problem for the country today is ____" with two choices to fill in the blank: "seeing discrimination where it does not exist" or "not seeing discrimination where it really does exist." The results confirmed the political chasm that we face. Republicans favored the answer "seeing discrimination where it does not exist" by 77% versus the 22% who chose the other option. Democrats were opposite with only 22% saying "seeing discrimination where it does not exist" and 77% indicating "not seeing discrimination where it really does exist." This goes back to the police killings of Black men and women in the last ten years. Viewed from liberals, these are clear examples of racism inherent in police departments. Viewed by conservatives, this is police protecting society from violent offenders. The truth is somewhere in between. In some cases, there is a clear indication that police opened fire without justification. In others, the evidence is less clear and who is to blame can't be answered with just the evidence that is available to the public, but that has not stopped both liberal and conservatives from rushing to conclusions and issuing judgments based on very little data.

That is the history and the opinions, but what are the facts and data? The FBI maintains the Law Enforcement Officers Killed and Assaulted (LEOKA) database, which reported forty-seven officers' felonious deaths in 2020, *felonious* meaning they were killed in a criminal act. The total of forty-seven is actually one less than 2019 and is less than average for the previous five years. Most of the deaths (twenty-five of forty-seven) were in the South, where the population is 40% of the overall country, but the amount of law enforcement officers which have died by criminal act averages 52%.

On the other side of the coin is the number of people killed by on-duty police officers. The *Washington Post* has compiled a database of all these deaths from 2015 until the present.[20] It lists all who have died by gunshot from law enforcement. Unfortunately individuals like George Floyd who died from an officer kneeling on his neck in 2020 or Freddie Gray who passed away in 2015 from injuries while being transported in a police van aren't listed. While some may question this, the cases where people died from jail suicide or physical injuries are a subset that doesn't have data readily available. The FBI has started a use-of-force database, but it is only voluntary participation and at the present less than half of law enforcement agencies are participating. Why the remaining agencies don't participate whether due to lack of resources or lack of interest is unknown.

While the *Washington Post* only gathered the data for individuals shot and killed by police, just from 2015 to 2020 there were 5,960 gunshot killings by law enforcement, almost 1,000 per year. The data indicates the name, date shot, what they were armed with, age, gender, race, location of the shooting, and if they had signs of mental illness. Analyzing the data shows some very interesting patterns. One surprising detail was that a white forty-one-year-old man was armed with an air conditioner—still not quite sure how that was done.

Not surprising is that most are male. Of the 5,960 deaths, 5,669 were male (95.1%). The data shows that 24% (1,421 of 5,960) were Black, while Black people only make up 12.8% of the population. This led the *Washington Post* to claim that "the rate at which Black Americans are killed by police is more than twice as high as the rate for White Americans." This is true, if you base it on overall population, but not if you base it on the percentage of those in poverty, as the following table shows.

Race	Percentage of Population	Percentage Shot by Law Enforcement	Percentage of People in poverty
White	54.1%	45.9%	41.6%
Black	12.8%	24.0%	23.8%
Latinos	17.9%	16.8%	28.1%
Asian	5.7%	1.6%	4.3%
Other	6.1%	2.2%	2.2%
Race Unknown	3.4%	9.5%	-

TABLE 4. INDIVIDUALS KILLED BY LAW ENFORCEMENT NATIONWIDE

When adjusted for poverty, the rate of Black people being shot by police is not out of alignment with the number of white people getting shot, and in fact by probability is less. White Americans living in poverty are actually getting shot more frequently, and with white people murdered by law enforcement totaling 2,723, the number is far larger. This also holds true with arrest rates. When you look at the FBI data, Black Americans are about 26% of arrests, while white people and Latinos are lumped together at 68%, which compares to their combined poverty rate of 69.7%. Race doesn't appear to affect getting shot by law enforcement as much as poverty does. And let's be honest: except for a few rare cases, it's not upper- or middle-income people getting shot and arrested. Criminal violence plagues low-income areas far more often. Does that justify this? No; 18.8% of all Black Americans live below the poverty line, as opposed to about 8% of white Americans. Black people are disproportionately affected by poverty, which impacts them in other ways, crime statistics just being one area.

And why are more Black Americans living in poverty? Well, the first point to consider is that they started with almost zero accumulated wealth in 1865 at the end of the Civil War and then suffered real and verifiable racism until the late 1960s. Many could argue that it is the discrimination and racism inherent within poverty that is reflected in the law enforcement killings.

But this can't really answer the question of whether there is systemic racism in policing if we assume that the majority of these people were actually committing a crime or, and the data shows this, trying to commit suicide by having law enforcement kill them. To determine if racism exists in police departments, we need to look at the unarmed people who are getting shot and killed. This is the heart of the Black Lives Matter protest that unarmed Black men are being gunned down by police. And what does the data say? Unarmed people shot and killed by police in the six years between 2015 and 2020 numbered 381. And while the data indicates that some of them were behaving aggressively, we are going to assume that since they were unarmed, there should have been a way to avoid killing them. And the same table for unarmed people shot and killed by law enforcement appears below.

Race	Percentage of Population	Percentage of unarmed people killed	Percentage of people in poverty
White	54.1%	42.2%	41.6%
Black	12.8%	34.1%	23.8%
Latino	17.9%	18.1%	28.1%
Asian	5.7%	2.1%	4.3%
Other	6.1%	2.6%	2.2%
Race Unknown	3.4%	1.6%	

TABLE 5. UNARMED INDIVIDUALS KILLED BY LAW ENFORCEMENT NATIONWIDE

So, does this prove systemic racism in policing? Looking at the results, Black people are more likely than other races to be killed when they are unarmed. Of the unarmed killings, 34.1% were Black victims, almost three times the overall percentage of Black people within the population, and one and a half times the percent of Black people within the people living in poverty. By the numbers, white people are killed in greater numbers than Black people, and it is very hard to label all police departments as having systemic racism based on the data above, especially as regions and states vary greatly. The table below shows the percentage of the US population per region and the percentage of the population in each region that is Black compared with the unarmed killing.

Region	Percentage of US population	Black population within region	Percentage of Unarmed killed	Unarmed Black people killed
South	40.0%	18.7%	171 (45%)	81 (47.4% of 171)
Midwest	20.8%	10.5%	67 (18%)	17 (25.4% of 67)
Northeast	17.1%	12.2%	23 (6%)	10 (43.5% of 23)
West	22.1%	4.8%	120 (31%)	22 (18.3% of 120)

Table 6. Unarmed individuals killed by Law Enforcement by Region

Of the Black population, 58% live in the South, but that does not explain why with only 18.7% of the overall population, Blacks are 47.4% of the unarmed killings by law enforcement, the highest percentage of any region. While in the West with only 4.8% of the region's population, Black people are killed 18.3% of the time, almost four times their percentage of the population. While the majority of the unarmed killings were of white people (158), all the regions show that Black men have a far higher likelihood of being shot and killed by law enforcement when unarmed, even when factoring in for poverty. And as was said earlier, some of the events have been clear overreactions by the police.

One reporter found that fifteen of the officers guilty of fatally shooting unarmed Black people had been involved in multiple shootings, while most police officers never fire their guns in the line of duty.[21] Nineteen of the officers were rookies with less than a year of experience, one with only four hours on the job. Overall, she found that 75% of the officers involved in the shootings of unarmed Black men were white. The article indicates that these weren't exactly our finest, but if so, why were they on duty? Small police departments in rural or poor areas are unable to compete with the salaries of wealthier areas, and they are having trouble getting qualified officers. Further, police unions have contracts that make firing police officers especially onerous. There have been consequences for these shootings. They have led to thirty judgments or settlements totaling more than $142 million, and dozens of lawsuits are pending.

There is something else that needs a little more discussion: we have increased the burden on our police officers. They are no longer just responsible for policing; we have added two other roles that massively complicate their lives. They are much more likely to encounter people with mental illness. Of the 5,960 law enforcements killings in the database, 1,387 cases (23.3%) indicated signs of mental illness in the data.

The second new responsibility is terrorism. Since the 9/11 attacks there have been twenty-three domestic attacks by radical Islamist extremists. These are far outnumbered by the incidents caused by domestic terrorist groups—sixty-two attacks. Even still, there is a clear perception of a threat from radical terrorists. The issue is that in terrorist situations, police officers have been increasingly trained to respond in an aggressive manner because that is what is required. Further, mass shootings in schools and churches also require an aggressive response from police. Between 2017 and 2019, eighteen police officers were ambushed out of the 195 officers killed. We live in a time of not only racism and social change, but also a time where more pressure has been placed on police.

This does not change the fact that Black Americans are more likely to live in poverty and more likely to be killed by law enforcement, even when unarmed. These two trends and our history make the case that racism is still present in America, even though progress has been made. So, what needs to be done?

Prejudice is a burden that confuses the past, threatens the future, and renders the present inaccessible.

—MAYA ANGELOU

The first thing to consider is the representation in the media. There have been multiple studies showing that Black Americans are far overrepresented in the news media when it comes to stories depicting crime or poverty. Even when by sheer numbers, white people commit more crimes and there are more white individuals living in poverty. While white poverty appears to be more focused in rural areas and Black poverty in inner cities, there is still enough white urban poor to question this unequal treatment.

Multiple studies have claimed that this bias is due to the management of the news media and feeding into fears from white people. The problem is there is very little hard evidence. No one has sat down and watched millions of hours of television to come up with a comprehensive study. The studies found in the research for this book all utilized small sets of data and a lot of opinion, which adds the potential for error. But all the studies did indicate the bias that minorities were depicted in poverty and crime more frequently. Remember those police officers' deaths earlier? They received a massive amount of media coverage, especially the ambushes that skew the public perception. Cases of police officers being ambushed in the last few years have been blamed on the Black Lives Matter movement, even when one was committed by a rightwing organization and others appear to be individuals acting on their own. But how do you act on this without stifling or even appearing to stifle free speech? The issue of fake news and alternate facts will be addressed in later chapters.

The second thing to consider is the make-up of police forces. There have been a number of incidents cited where comparisons of the racial profile of a population doesn't match the racial profile of the police department. Ferguson, Missouri, is an extreme example. When Michael Brown was fatally shot on August 9, 2014, the population of the city was about 67.4% Black out of a total of about 21,000 during the 2010 census. The 1990 census had the city as 22,000 people of which 73.8% were white and 25.1% were Black. Within twenty years, the city had undergone a dramatic shift in race, but the police force had not. In 2014, it had fifty white police officers and only three Black officers. That racial disparity between the police department and the community is something that needs to be tracked and reported. Not for some type of quota system, but rather as a flag that when unarmed killings or excessive use of force is detected that it would bring additional scrutiny. While some will argue that quotas or guidelines are needed, the counterpoint is that Federal government oversight of the composition of the local police forces would likely increase tension. It could promote changing the racial profile at the cost of losing experienced police officers who had shown no indication of racism. While unqualified officers have been hired, some of those unqualified officers involved in multiple shootings of unarmed Black people were themselves Black. So, there is no guarantee that a quota would be filled by qualified officers. But tracking the racial profile and notifying police departments that it would be flagged for any racial incidents puts a subtle pressure on those departments to change, which hopefully will achieve better results than an active federal involvement.

This should also be extended into a national database of law enforcement officers that would record and track shootings, disciplinary actions, and criminal activity by officers. This could then be used as a means for checking backgrounds to prevent unqualified officers from bouncing from jurisdiction to jurisdiction trying to keep ahead of their past failings.

Reparations is a subject that comes up and is sometimes tied back to the "forty acres and a mule" promise made to freed Black slaves that was never fulfilled. "Forty acres and a mule" was an agreement between General Sherman, Secretary of War Edwin Stanton, and Black ministers in South Carolina in 1865. Sherman issued a military order that plantations along the South Carolina–to-Florida coast would be split up and given to freed slaves. The mule was added in later negotiations. Breaking up and allotting ex-plantations to formerly enslaved people gave them land and a means of income. It was a simple solution with minimal cost to the Federal government. The problem was that Lincoln's successor, President Andrew Johnson, was swayed by Southern planters and rescinded the order, returning the land to the original owners. This is one of those heartbreaking moments in history where we can argue about morals, reasons, and alternative outcomes over a decision in 1865, but we can't change what has already happened.

Which brings us to the present-day. If Black Americans are to get some form of reparation, money to compensate for the wrongs of slavery and Jim Crow, who gets it and how much? The *who* is tricky. Do Black Americans whose ancestors emigrated from Africa or the Caribbean after the Civil War get it? What about people who were free prior to the Civil War? How do you address people of mixed lineage, descended from some free and some slave ancestry? People of mixed race? Many Black Americans have succeeded and prospered; do they still get a reparation even if they are worth millions? And how much? The equivalent of forty acres and a mule in modern dollars? The problem with such an idea is defining what form the reparations take and how much. Any financing by the Federal government would surely end up as deficit financing, only adding to the national debt, which is owed by everyone. There are many issues with this and lots of questions come up. Do we end affirmative action laws and other programs that presently exist if we provide reparations? There doesn't seem to be a feasible way to provide reparations. One potential would be federally backed low-interest loans for Black people to buy houses or businesses, but that would only benefit those not living in poverty with the means to start a business or pay for a house. I welcome any proposals for a solution, but I think the right answer lies elsewhere.

To bring about change, you must not be afraid to take the first step. We will fail when we fail to try.

—ROSA PARKS

The right answer may be to continue down the path we are presently on with a few changes. There has been improvement, and while many have wished for more and sooner, societies, like large ships at sea, don't change course rapidly. But there are issues that need to be addressed; the foremost one is poverty. Addressing poverty is extremely hard for a national government. Throwing money at poor people seems like an obvious answer, but too often they may lack the skills to invest, to understand where to spend their money to make a long-term improvement. Windfalls to the poor sometimes produce minimal results. A fact that has to be faced for all people of every race living below the poverty line is that there are people who won't make the effort to help themselves. That is a concept that may make people uneasy, but past studies and results do bear this out.

The solution is to set up programs that can help poor people to climb out of poverty. It is far easier to help a person climb out of a hole than to try and pull an uncooperative person out of the hole that poverty represents. The first part of the solution must be education that will be addressed in subsequent chapters. The second is to match up jobs with the poor, and that will be addressed in the chapter on housing. The solution to poverty must start with people wanting to climb out of poverty and ensuring that the ladder exists to help them.

Again, the house of America has improved from earlier times. It looks better than many of our neighbors, but it still needs a lot of work, so what do you do? You can't go bankrupt to create a perfect house no matter how much you or the American family want to, but you need to perform all the repairs you can afford to make the house as peaceful and fair as it can be and ask for tolerance from those still trying to improve and from those suffering from racism.

There is no single change that needs to be made to address racism specifically. America has made progress and needs to continue to try to decrease racism. Based on the discussion in this chapter, I recommend the following policy changes:

- Encourage all law enforcement agencies to participate in the use-of-force database supervised by the FBI as part of the Department of Justice.
- Participate in a national database of police officers for the sharing of employment history, training, and incidents.
- Have police agencies report their racial profile with the known racial profile of the community they serve.

These three items would be a prerequisite for any agency to receive Federal funding. It is only with knowledge that we can identify racism to then be able to take steps to correct it.

This was not an easy chapter to write. As a white American raised by Midwestern middle-class parents in the '60s and '70s, I have been slowly coming to understand the extent of the hidden racism that still exists. The search for some kind of hard data to prove or deny police racism was difficult, as no one really tracks the data. When I stumbled onto the *Washington Post*'s efforts, I found the data needed for the analysis.

A final note: when I was conducting a final review of the data, I rediscovered the case of Philando Castile who was shot and killed by police in July 2016. I remembered the case clearly and how to me it was one of the cases where I thought police killed an innocent Black man. It got me started questioning the police killings. I went back and checked and found that in the *Washington Post* data, his status was marked as "armed." He had a licensed gun in his car at the time of the shooting, so he had been lumped in with all the shootings where the victims had been armed, even though he was licensed to carry a concealed weapon and had not been reaching for it. I almost changed my spreadsheet to unarmed which would then change the number of unarmed killings and the percentages of everything that flowed from that one cell on a spreadsheet. But you have to respect the data and keep it in alignment with the source data you started with. Letting the data take the book where it needed to go has been the overriding principle of this book, but I felt that Philando Castile needed an apology from me for not putting him with the unarmed victims.

11.
Immigration

I had always hoped that this land might become a safe and agreeable asylum to the virtuous and persecuted part of mankind, to whatever nation they might belong.

—George Washington

The third hot button issue that has been very divisive within American society is immigration. For a country that was built by the massive immigration of the late 1800s and early 1900s, this is truly ironic. The images of "caravans" of illegal immigrants walking to the southern border has been used to provoke fear and justify the building of a wall. But one of the images that sticks with me is an illegal immigrant pushing her daughter over the new taller Trump wall. The child broke her legs when she fell down on the American side. What wall can hold up against that level of desperation to cross it? Especially considering these refugees have walked over a thousand miles, through extreme terrain, and we think a wall is going to stop them in the last half mile.

People may say that good fences make good neighbors, but I have never seen a fence stop somebody desperate enough to cross it. Historically, the Great Wall of China, the Maginot Line of France, and Hitler's Atlantic Wall of World War II all failed to keep out the people they were trying to hold back. The idea of a wall represents a simplistic solution. Especially when the terrain and obstacles along the southern border are considered. The problem is that walls haven't stopped people from coming or from crossing the walls. What would it take to stop all these illegal immigrants from crossing the southern border—minefields and soldiers with orders to shoot to kill? A few radicals will embrace that idea, but the cost of dead bodies piling up on the border, especially of children, would be a stain on America's soul and something our enemies would use to pillory us. And rightfully so. Think of the countries where people are shot crossing a border; North Korea springs to mind as an obvious example. That is the wrong solution for America.

The slaughter of illegal immigrants is not an option, and a wall is ineffective, so what is left? The present option of arresting and then exporting people back to their country (or somewhere else) has been used not only in the last few years, but far longer. "Catch and release" has been going on for a long time, and besides the cost of the program, the other issue is you don't catch everyone, so it still leaves the question of what to do with those that you missed. And even more importantly, what do you do with all the millions of undocumented immigrants who have been living hidden in the country for years? What about the legal immigrants and the backlog of people applying for green cards? We are at levels far above what the Rockefeller Commission recommended in 1972 to curtail the population explosion that the commission felt would endanger the quality of American life.

The period from 1850 to 1930 was the heyday for immigration with millions of migrants coming to the United States, which had an extremely low population density of about seven people per square mile in 1850. Now, that assumes the 1850 population of 23 million was scattered across the same area as the present United States, but most areas were US territories. That availability of land coupled with adverse conditions in Europe, such as high population density, the Great Potato Famine of Ireland, and repressive governments combined to create an impetus for people to leave. By 1930, the US population was 123 million and population density had risen to forty-one people per square mile. With population increasing and the need to curtail immigration, Congress began passing laws to restrict immigration, and some of these laws had very strong racist overtones. The Immigration Act of 1924 was aimed at restricting immigrants from Southern and Eastern Europe and consolidated the prohibition of Asian immigrants that had started in the Chinese Exclusion Act of 1882.

Our present population of over 334 million means that the national average population density is now 112 people per square mile. And the population density varies greatly from state to state with New Jersey at 1,218 people per square mile down to Alaska at one person per square mile. But the initial reason for large-scale migration—low population—no longer applies. "Give me your tired, your poor, your huddled masses yearning to breathe free," was written in 1883 by the poet Emma Lazarus when the golden age of immigration was upon us. We cling to that quote believing that it defines America, when the reason and needs that drove it have fundamentally changed. The reality today is that America is faced with many things, but two core issues are poverty and massive national debt, neither of which is helped with a large influx of poor people, legal or not.

The present illegal immigration problem breaks into two groups: the approximately 12 million people presently in the United States, of which over 8 million have been here for more than ten years, and the over one million illegal immigrants who will either attempt to cross the border or overstay their lawful visas.

We will focus first on the people attempting to cross the southern border. They comprise the majority of attempts and generate the most emotional response by the public. Arrests of illegal immigrants were decreasing; the highest levels were seen between 1983 and 2006 when more than a million people per year were caught attempting to cross the border. These were predominantly single Mexican men seeking out work and sending money back to families in Mexico. In recent years, the average has been about 500,000 per year, but in 2019, it spiked up to 1.1 million. A change in demographics has also occurred with more immigrants from the Northern Triangle of Central America, El Salvador, Honduras, and Guatemala and many more are families or even unaccompanied minors. In 2019, a little over 250,000 were from Mexico, while more than 650,000 were from the Northern Triangle. This large flow from there doesn't try to cross illegally, they try to claim asylum. There are lots of reports on the corruption, poverty, overpopulation, and chronic violence in the Northern Triangle that is providing the impetus to drive these people north. There are further reports on the violence, criminality, and hardships that people travelling from the Northern Triangle through Mexico to the US border suffer through, but we need to understand that isn't something that the United States is responsible for nor can affect. That may seem harsh, but every individual made the decision to attempt the journey and as one of the core tenets is personal responsibility, that decision is not one we can be responsible for.

The Trump administration had implemented programs to make crossing the border as difficult as possible by telling people to wait in Mexico while their asylum claim was being reviewed under Title 42 due to the pandemic. There was also a policy of separation of families at the border, which was an attempt to discourage people from trying to enter the States. Based on the numbers, that was unsuccessful at keeping new migrants out. President Biden's policy was to continue Title 42, which prohibits entry for medical reasons, but they were admitting unaccompanied minors, and interviews with migrants in Mexico indicated that many had tried to send their children over alone when the adults were blocked. With COVID measures expiring, Biden has moved to limit asylum seekers and have them apply electronically from afar, rather than admit more to the United States.

The United States signed the United Nations 1967 Protocol Relating to the Status of Refugees, which requires the country to provide protection to those who qualify as refugees or claim asylum at the border. This was codified into US law, and while many want to blame presidents, the reality is that they are only enforcing the law that requires applications of asylum to be reviewed at the border. The problem is that of those trying for asylum, federal agents deemed that some had claims of "credible fear." The number of people identified as having "credible fear" and thereby a possibility of getting asylum has skyrocketed from 5,523 in 2009 to 102,204 in 2019. That means that these people may have valid means to claim asylum and can't be turned away at the southern border. The backlog of these asylum cases is enormous. The Transactional Records Access Clearinghouse (TRAC) at Syracuse University tracks immigration statistics and indicates that immigration courts have over 2 million cases in backlog as of 2023, up from 1.2 million in January 2021. Of these applicants, for 2019 fewer than 80,000 were granted asylum or refugee status, and they were predominantly from Congo (12,958), China (7,478), Venezuela (6,821), Burma (4,932), and Ukraine (4,451). In 2019, the three countries of the Northern Triangle had a combined total of less than 8,000 asylum requests approved.

The problem is that while the potential for asylum at the southern border exists, people from the Northern Triangle will continue to try and get into the United States even though the likelihood of them gaining permanent asylum is extremely low. The data and the low chance of success are not dissuading people, but success stories are giving them hope, which can be seen by the numbers. Even separating families and making people wait in Mexico has not been a proven deterrent. So, what is to be done?

No wall or border patrol is going to stem the tide as long as there is bitter poverty, corruption, and violence providing motivation, and people believe they have a chance of entering the United States either illegally or by asylum. Past US administrations had provided economic assistance to the Northern Triangle to lower the motivation. The Trump administration held that assistance back demanding that the governments of the Northern Triangle block the migrants. That didn't happen, and now numbers have increased beyond what was going on in the Obama years. But even reestablishing economic assistance won't change the motivation. The pattern of travel to the United States is established, and assistance won't stop the migration, and how much it mitigates the desire to migrate will be impacted by how much assistance actually reaches the poor people choosing to migrate. The United States cannot afford to have a million poor migrants crossing the border expecting to settle in the country. With our present poverty and national debt, it just cannot be handled at this time. It is unfortunate, but we need to remove the hope of getting in.

The plan for the southern border needs to be to increase the deterrent. First, simply ban all asylum claims for countries that do not border the United States. The Protocol Relating to the Status of Refugees defines a refugee as "owing to well-founded fear of being persecuted for reasons of race, religion, nationality, membership of a particular social group, or political opinion." The people from the Northern Triangle who are applying for asylum at the southern border fear crime and poverty, and they had to pass through other countries to reach the United States. We can retain the option for people to apply for asylum at a US Embassy, but not at the border unless they are from a country that borders the US.

Second, reject all pending immigration asylum cases that do not meet the above criteria and send the people in the United States either across the border to Mexico or on a plane to their home country. The almost 2 million backlog of cases need to be simply closed. This action may well escalate to the Supreme Court, but asylum seekers can't have their lives in limbo, and the United States can't afford to have all those pending cases, especially considering the exponential growth in cases.

Third, anyone caught crossing the border illegally is photographed, fingerprinted, and returned across the border with the warning that if they are caught again, they will be sent to jail. The second time is jail and manual labor before being returned across the border. The concept of manual labor may seem abhorrent to many, but the idea is people caught a second time will have to do trash pickup, clean up, or physical labor to upkeep the walls—anything to step up the deterrent. The counterpoint is to reestablish the economic assistance to the Northern Triangle, to try and reduce the motivation to emigrate. The total assistance to the three countries was less than a billion dollars during the height of the Obama spending on foreign aid, and spending the same amount or even half a billion should lessen migration. That assistance will take time to reduce the motivation so it cannot be the only solution.

Many will see this as harsh and point out that very few of these people are violent, but they are trying to enter illegally, and America cannot handle a million poor immigrants when we are already struggling with over 40 million in poverty right now. Also, by accepting displaced poor from these countries, we are reducing the incentive for those countries to deal with their problems. If people who are dissatisfied with the country's problems can just leave for the United States, then those same people are not working to correct the country's problems. The United States acts like a pressure relief for Mexico and the Northern Triangle, removing the potential for change where it is needed. We can continue to have applications for asylum be made at US Embassies or Consulates, but not at the border with Mexico.

The next question is what to do with the 12 million undocumented people living in America. This is different from the people actually trying to cross. They are already in the United States, and 8 million of them have been here for more than ten years. While there have been claims that undocumented residents are responsible for crime and a drain on social services, studies show that the vast majority have not committed any crimes except immigration and have built and sustained lives within the country. A Texas study published in December 2020 indicated that undocumented immigrants were almost a third less likely to commit crime as US born citizens or legal immigrants.[22] Texas not only collects data on arrests and convictions, but it is the only state that collects data on the immigration status of the individual. Researchers found that 96% of the entries indicated whether the person was undocumented, a legal immigrant, or native born. The table below summarizes the data.

	Arrests per 100,000 persons		
	US-born citizens	Legal Immigrant	Undocumented Immigrant
Violent Crime	213.0	185.3	96.2
Property Crime	165.2	96.2	38.5
Drug Violations	337.2	235.6	136.0
Traffic Violations	68.3	86.7	38.1

TABLE 7. DOCUMENTATION STATUS VERSUS CRIME IN TEXAS

While this data is limited to Texas, it is confirmed by other studies indicating that undocumented people pay taxes and provide a benefit to the economy. The view that undocumented people are a source of problems is only supported by the opinions offered to support building the wall and deporting all undocumented people. Studies on the almost million people in the Deferred Action for Childhood Arrivals (DACA) program have indicated that not only has DACA improved the lives of its recipients and their families but provided economic benefit as well. A study by Ike Brannon and Logan Albright published by the Cato Institute in 2017 indicated that deporting 750,000 people would have an immediate cost of $60 billion and a long-term cost to the economy of $280 billion over the following decade.[23]

This should be no surprise. To qualify for DACA, recipients had to meet several requirements. They had to show continuous presence in the United States since 2007, arrived in the United States before age sixteen, be at least fifteen years old at the time of the first application, be enrolled in or graduated from high school, and have no convictions for a felony, significant misdemeanor, or three or more minor misdemeanors. By applying for DACA relief, applicants trusted the US government with numerous personal details, including their residential history, school location, and workplace, and they paid an application fee of $495, including biometrics, a steep price for many applicants and their families. Recipients were required to renew their work permits every two years to maintain their ability to stay in the country without fearing deportation and to work legally. DACA individuals have achieved English proficiency (87%), and most have graduated from high school (98.6%). Though they are undocumented, DACA recipients are contributing to America in attempt to realize the American dream. To deport them now makes no sense, and to deny them a path to citizenship makes a cruel irony of the effort they have done.

Conservatives can wail about amnesty, but they need to remember that President Ronald Reagan provided a means for the undocumented to come out of the shadows in 1986, and approximately 3 million took advantage of this. The 1986 effort required that applicants had been in the country prior to 1982, had not had any criminal offenses, had a minimal knowledge of English, and paid a fine. The program was a success, but there were many millions who did not take the offer and remained undocumented. Similar ideas have been proposed in recent years, but none have passed due to these programs being characterized as amnesty.

But this is not a blanket amnesty, but a clear set of requirements for people who are willing and able to move from undocumented status to documented and forge a path toward citizenship. The plan should be to move the DACA recipients to that path without further delay. They have already met the requirements. Second, apply the same criteria except for the age requirement to the remaining undocumented people. They need to show that they have been living in the States for at least five years continuously, have no convictions for a felony or significant misdemeanor, be able to speak and understand English, are willing to pay an application fee of $500, and provide biometrics and residential history. The ability to understand English could be waived or delayed, but it needs to be shown that the person is attempting to assimilate. This should also be extended to the children of any adults who qualify, so we don't create further problems down the road.

The path to citizenship should be the same as presently established by US Citizenship and Immigration Services, that after being awarded lawful permanent resident status, the person demonstrates continuous residence for another five years, and then meets all the requirements for being awarded citizenship.

The flip side is that after this, anyone caught in the United States and confirmed as undocumented gets deported without question. Cities and jurisdictions that have been claiming "sanctuary status" need to stop after this attempt to bring everyone who can and wants to achieve documented status. We all need to agree that if you haven't come in from the shadows, there are no more second chances. Liberals may find that a bitter pill to swallow, but it must be the agreement that after the Dreamers, their families, and anyone who wants to become documented, we cannot continue to not only protect but encourage illegal immigration. The problem of poverty in America needs to be addressed, and it can't be if we continue to add to the problem with illegal immigration.

The last part of immigration is legal immigration and naturalization. Legal immigration is where people are granted lawful permanent residence (LPR), i.e., a green card, and are then allowed to live and work in the United States. Naturalization is the process where those with LPR can gain citizenship after being in the country for five years. There are also visas that allow temporary workers to enter the country, H1B visas are specialty occupations, H2A are for agricultural workers, and H2B are non-agricultural workers. The visas are temporary and need to be renewed, but they can also be converted into permanent green cards.

We have averaged about 750,000 naturalizations per year for the last six years. That makes sense when compared to the over one million people granted lawful permanent residence. We have averaged granting LPR to about a million people since 2000. Even the Trump administration held that average. In the 1970s, the United States averaged about 425,000 people granted LPR. The bulk of the grants are to spouses of US citizens (25.7%) and their immediate families (40%). Only about 13% are to people for work visas, and they usually need to be sponsored by their employers. The remaining is split between refugees (10.6%), asylum seekers (3%), Iraqis/Afghans employed by the United States (1%), victims of human trafficking and other crimes (2%), and diversity (4.5%). Diversity is the visa lottery where anyone can get in from a country that has low immigration rates from the previous categories. At present, there is a backlog of over five million applicants for green cards. Over one million are for the employment-based green cards.

The 1972 Rockefeller report identified reducing legal immigration as a need to limit overpopulation in the United States. Climate change is not just about reducing CO^2 being released; it needs to be about reducing the amount of garbage generated, the human-related impact to the environment. That must be about reducing population growth. Further, the plan to bring as many undocumented people as possible into legal residence will increase the demand on family-sponsored applicants, which already has an unbelievable backlog.

The other problem is the limits on the amount of employee-based grants of LPR. At present, the limit is 140,000, but the high-tech industry and several others want to increase the amount of H1B employment visas and green cards to raise the number of skilled workers they can bring in. But allowing industry to bring in outside workers means there is less pressure on industry to develop American talent in conjunction with American universities. This is a tradeoff between getting the required skills locally or from overseas.

The policy should be to enact limits on lawful permanent residence to keep America's population from growing too much and to put pressure on employers to not bring in foreign talent but to work with the education system to develop American talent. But where to put those limits? This is especially difficult as there will be a higher need for skilled high-tech people to help move from fossil fuels to a renewable-energy economy. Is local talent sufficient? America has always had a melting pot culture with a good deal of diversity in its cities. Can we impose limits to legal immigration and not negatively impact that? Further, limiting refugees and asylum seekers runs contrary to the very concept that George Washington expressed at the beginning of the chapter. Lowering limits on legal immigration is best tabled at this time while America struggles with some of the other issues in this book. It should be revisited after five or ten years to see what needs to be done. That will both disappoint conservatives who are convinced that limits are too high and liberals who feel they are too low.

The final question on immigration is about "anchor babies." The Fourteenth Amendment ratified in 1868 reads: "All persons born or naturalized in the United States, and subject to the jurisdiction thereof, are citizens of the United States and of the state wherein they reside." The question comes into how this applies to children of people born here, but not to citizens. When it was written in 1866, it was to establish citizenship for the freed slaves from the Civil War. The problem is how does it apply to children of illegal immigrants? The Supreme Court in 1898 held that a child born to foreigners "but [with] a permanent domicile and residence in the United

States, and are there carrying on business, and are not employed in any diplomatic or official capacity under [a foreign government]" did qualify under the Fourteenth Amendment to receive US citizenship.[24] This has not been taken any further to determine if it applies to parents illegally present in the United States, but it has always been viewed that a child born domestically to parents who are legally in the country is a US citizen.

But this precedent has been abused. There is even a business where pregnant women are legally visiting the United States to have their child born here to establish that child as a citizen. Estimates are about 36,000 children a year are born in the United States to legally visiting parents, who intend to get citizenship for that child. The Trump administration issued revised guidelines barring pregnant women from being granted legal visas. The question is what should the policy be and how do we address this?

The issue is the way the Fourteenth Amendment was written—with very little detail. The best solution is a Constitutional amendment to revise the Fourteenth Amendment to read "All persons born **of at least one citizen or legal permanent resident of the United States** or naturalized in the United States . . ." which would remove any doubt.

12.
Guns

With the right to bear arms comes a great responsibility to use caution and common sense on handgun purchases.

—Ronald Reagan

Ten years after he had been shot by a would-be assassin, Ronald Reagan issued the quote above in support of the Brady Bill, which proposed background checks and restrictions on gun sales. He was still a member of the National Rifle Association (NRA), still owned guns, but apparently getting shot had made Reagan a little more open to gun control.

This is the fourth hot-button issue: guns and the right to own arms as defined in the Second Amendment. It reads, "A well-regulated Militia, being necessary to the security of a free State, the right of the people to keep and bear Arms shall not be infringed." So, let's understand the 1789 meaning based on the military hardware available at the time. Private citizens could own and keep shotguns, rifles, and pistols. Some say it meant that the Founding Fathers thought that people should not be allowed to own cannons or mortars. There is nothing in the Constitution or anything written by the Founding Fathers that prevents people from owning cannons or mortars. The Constitution even allows Congress to issue "letters of marque and reprisal" that would have allowed private citizens to use armed ships against countries the United States was at war with. The Supreme Court later decided in a ruling in 2016 that "the Second Amendment extends, prima facie, to all instruments that constitute bearable arms, even those that were not in existence at the time of the founding, and that this Second Amendment right is fully applicable to the States." This opens the door to private citizens owning any weapon that can be carried by a single person. There are even provisions for people to own cannons, mortars, and machine guns, but only with special approvals from the Bureau of Alcohol, Tobacco, Firearms, and Explosives (ATF) and their state.

And how many guns are in America? The Small Arms Survey of 2017 estimated that almost 400 million firearms were in civilian possession in the United States; that is more guns than people. Other studies have put the number higher, but for this discussion we will use 400 million. At 120 guns per hundred people, even the "small" number of 400 million guns places the United States far ahead of any other nation for guns in civilian hands. At just over forty-one guns per hundred people, Switzerland is the next closest, and there are a number of countries with just above thirty per hundred, but none of them come close to the United States. The figure of 400 million may seem outrageous, but consider that the ATF tracks gun manufacturing in America, and in 2016 alone, US gun manufacturers produced 11.5 million guns. They exported fewer than 400,000 and imported almost 4.5 million for an addition to America's gun total of 15.6 million. That was just in 2016, which was a banner year, but from 2009 to 2018, the US gun total increased by an average of 12 million guns each year. These totals don't account for the weapons that federal law requires to be registered with the ATF, such as the almost 640,000 machineguns, 150,000 sawed off shotguns and almost 1.5 million silencers. Those are not included in the 400 million guns total either.

But what are the consequences of all those guns in America? The Center for Disease Control and Prevention records the number of gun deaths in America. For 2020, it was 43,595, of which 24,245 (56%) were suicides. Gun control advocates have pointed to suicide as one reason for gun control, but then, if a gun wasn't available, would they try something else? The point from the advocates is that guns are very effective at suicide. Using a gun for suicide is fatal in 82.5% of attempts; the next closest is drowning/ submersion, which is successful 65.9% of the time.[25] But that misses the point. Removing guns may not stop the attempt, and while it may make the attempt less likely to succeed, is that something that justifies gun control? How much gun control would be required to impact the numbers?

The United States has 7.3 gun suicides per 100,000 citizens. While that number is the highest in the world, there are a number of countries with levels of gun suicides above two per 100,000 citizens, which have gun ownership levels of thirteen to forty-one guns per hundred citizens. A number of these are in Europe. Switzerland has 41.2 guns per hundred citizens and a gun suicide rate of 2.32 per 100,000 citizens. An effort to remove over 66% of America's guns just to reduce the number of gun suicides doesn't make sense when you consider that in 2016, Europe's overall suicide rate was 15.4 per 100,000 citizens versus the United States at 13.7 per 100,000 citizens. This has changed for 2019 with the United States at 14.4 suicides per 100,000, and Europe's average dropped to 10.5. But the point is that taking away the option of guns doesn't prevent people from using other means. Europe has a very low gun suicide rate compared to the United States, but overall has a comparable suicide rate. In 2019, Belgium had 13.9 overall suicides per 100,000, very close to the United States at 14.4 suicides that year, but Belgium only had 1.09 gun suicides per 100,000 citizens and has low gun ownership at 6.9 guns per hundred citizens (a mere fraction of US ownership). The conclusion is that the decision to attempt suicide is independent of whether guns are present. Access to guns provides the potential means that will be used. The focus should be addressing poverty and mental illness, which are the causes of suicide rather than the means that people opt for to attempt suicide.

The reason gun control advocates want to include gun suicides among gun deaths is that if you remove them, you are only dealing with 19,350 gun deaths for 2020, still a large number, but much fewer than 45,000 which includes gun suicides. Gun homicides are not one of the major causes of death in the United States at about 0.5% of all US deaths, including medical, accidental, and felonious. Now there has been a lot of discussion about how this compares with Europe and elsewhere. Japan has almost no gun homicides, and the United Kingdom has 0.02 gun homicides per 100,000 citizens, although the British Murder Mystery genre would make you think otherwise. Most European countries are between 0.06 (Germany) and 0.29

(Italy) gun homicides per 100,000 citizens, while the United States in 2019 was 4.6 per 100,000 citizens. That number of 4.6 puts the United States at a far higher rate of gun homicides than any other country that is part of the modern economy. You must look at the Philippines (7.62), Mexico (16.5), Jamaica (38.2), Honduras (28.65), South Africa (12.92), Venezuela (26.48), and a number of others for higher numbers of gun homicides per person. That doesn't put the United States in good company, so would disarming make a difference?

Looking at the countries with higher gun homicide rates, the United States has 120 guns per hundred citizens, and the countries with the higher homicide rates ranged between four to thirty-two guns per hundred citizens. It does not correlate to the number of guns or the number of gun homicides, but it does correlate to poverty. All the countries except one have levels of poverty in excess of the United States. The tables of countries and the analysis in the Appendix show the patterns between gun homicides, suicides, and poverty. The summary of the Appendix is that unless you massively reduce levels of gun ownership, you can't impact the gun homicides because the more important factor with gun homicides is poverty. European countries with high levels of gun ownership (35–42 guns per hundred citizens) don't have the same levels of gun homicides when their poverty levels are less than America's. Even within the United States, among the twenty-five states with the worst gun death rates, only three have a gross domestic product per citizen above the national average, and all three have low homicide rates, but high suicide rates. Thirteen of them have the worst GDP per citizen for the whole country. People will point to California and make the case that their low gun homicides are a result of California's strict gun laws. I would point to California's high GDP per citizen as the criterion that sets it apart.

This point is strengthened by a report released by the Center for Disease Control in May 2022 that looked at the increase in gun homicides between 2019 and 2020.[26] In 2020, with COVID raging in the background, gun homicides increased from 4.6 per 100,000 citizens to 6.1 deaths per

100,000 citizens, a 35% increase. The CDC looked at why this occurred and found that "rates of firearm homicide were lowest and increased least at the lowest poverty level and were higher and showed larger increases at higher poverty levels." The reality is that the volume of guns is not the driving factor in gun homicides; it is poverty that drives the crime rate.

No amount of partial disarmament is likely to affect it. Removing 66% of America's guns to lower us to a level with Switzerland means taking a minimum of 264 million guns out of private hands. Remember we are working with the "small" number of 400 million guns in America. Any idea of the government taking that many or more guns from people is going to be next to impossible. Not only because the Second Amendment bars the government from doing it, but also most Americans oppose disarmament and logistically confiscating 264 million or more guns is simply not feasible. This has not stopped politicians and the NRA from claiming that "the liberals are coming for your guns," which is a way of driving donations to the NRA and votes to the NRA's candidates. So, let's stop and agree that disarming America is not constitutional nor possible, and likely disastrous for a liberal government to attempt. No matter how often liberals may point to other countries and compare how much less gun violence they have compared to America, this is just not feasible. First, disarmament won't produce lower overall suicide rates, and poverty is the driving factor for gun homicides. Second, because most people legally own and use the guns. If 45,000 guns are used in suicides and homicides per year, that is 0.01% of all the guns in America. So, when we discuss disarming, we are really discussing punishing 99.99% of Americans for the actions of an incredibly small few. This is why the idea of disarming America is not only impossible and ridiculous, but it also blocks discussion of real gun control issues that need to be addressed.

The issue of mass shootings is the other point brought up by gun control advocates as a reason for more extensive gun control. On May 26, 2022, a list of the thirty worst mass shootings in the United States was published.[27] Half of them (fifteen) had been in the ten years preceding this book. At the time, the tenth worst and oldest on the list occurred in 1966 and involved a shooter who killed his wife and mother and then went to the observation deck of the University of Texas's clock tower. He killed three people on the observation deck and then used a bolt action rifle and killed another eleven and wounded thirty-one others. For almost two hours, he terrorized the University of Texas campus before two Austin police officers were able to reach the observation deck and, in a shoot-out, killed him. A friend once made the brazen remark that it takes some skill to kill that many with a bolt action rifle, but any damn fool can beat that number by walking into a crowded area with an automatic weapon—a crass comment but unfortunately very true.

The worst mass shooting in US history was in 2017 when a man opened fire from the top of a Las Vegas hotel into the crowd at the Route 91 Harvest music festival 450 yards from the hotel, killing fifty-eight and wounding 413 over the span of ten minutes. He fired over a thousand rounds of ammunition from fourteen AR-15 semiautomatic rifles, all equipped with bump-stocks that turned them into fully automatic military-equivalent weapons. Some were equipped with hundred-round drum magazines. He committed suicide as police were closing in. Semiautomatic rifles with multiple large-capacity magazines were used in seven of the ten worst mass shootings as well as all five of the ones that were in the last five years. These weapons have become the weapon of choice for mass murder. Thinking about banning them? It may be too late; gun manufacturers track the production of "modern sporting rifles," which in gun speak is the civilian version of military rifles, and they estimate the number already sold in the United States is over 18 million. That is likely an underestimation of the actual number in circulation.

Thinking about trying to do a buy-back or some other plan to get those weapons off the streets? In the wake of the Las Vegas shooting, bump stocks were outlawed, and the Bureau of Alcohol, Tobacco, Firearms, and Explosives issued a final rule that indicated "current possessors of bump-stock-type devices must divest themselves of possession as of the effective date of the final rule (March 26, 2019)." The ATF has collected less than 1,000 of the bump stocks sold. How many were sold? That answer is unclear, but it is greater than 200,000. There were several lawsuits filed by multiple organizations challenging the ban. In January 2023, the Fifth US Circuit Court of Appeals ruled that the law failed to define *bump stocks* as machine guns and as such the ATF did not have jurisdiction over them, which would kick the law back to Congress or the Supreme Court. Even though in another case, the DC Circuit Court of Appeals ruled that bump stocks converted semiautomatic weapons to fully automatic making them machine guns and able to be regulated by the ATF. Other lawsuits are pending. One lawsuit filed in federal court seeks compensation for individuals deprived of their bump stocks. And while some may be waiting to see how the lawsuits resolve, even if all the lawsuits opposing the ban were rejected, confiscation would likely prove ineffective.

One other thing: you don't need a bump stock to convert a civilian version of a military rifle into a fully automatic military version. There are multiple YouTube videos and information sources on the internet that show you how to convert them if you have the tools and enough metal-working skills to give it a try. While this is not the same as Prohibition when America tried to outlaw alcohol, it has to be seen as an example of not being able to put the genie back in the bottle. Realistically disarming America is not an option, but it has to be asked: why do American citizens need semiautomatic high-velocity weapons with large magazines? These weapons are a "force multiplier" allowing a single person to inflict massive casualties in crowded locations. There are several factors that need to be discussed.

First is the National Rifle Association, which was founded in 1871 with the intent of teaching gun safety and marksmanship. It wasn't until 1934 that it began to defend gun ownership rights. In 1976, the NRA established the NRA Political Victory Fund, a political action committee for lobbying and supporting pro-gun candidates. The NRA today identifies itself as "one of the largest and best-funded lobbying organizations." It receives large donations from gun manufacturers and is opposed to any form of gun control or limitation, helping to effectively kill the Brady Bill when it came up for renewal in 2004. So, the NRA has worked tirelessly with monetary support from gun manufacturers to make semiautomatic versions of military guns available to the public. They have also worked to block background checks and have prevented gun sales from being regulated.

The second issue is those calls by politicians and the NRA that "the liberals are coming for your guns." In an official 1995 NRA Institute of Legislative Action fundraising letter signed by then executive vice president Wayne LaPierre, they referred to federal ATF agents as "jack-booted government thugs." The NRA later apologized for the letter's language. This and government standoffs and fatalities at Ruby Ridge (1992) and Waco (1993) gave some groups the idea that they needed to be armed with military-grade weapons to confront government overreach. Though these anti-government groups declined from 800 groups in 1996 to only 150 in 2000, they have been increasing recently as evidenced by the events at the Capitol on January 6, 2021. Protestors from these groups have been openly carrying semiautomatic weapons around state capitols, which they are legally allowed to do. Some of the killings during the racial protests in 2020 and 2021 involved people openly carrying semiautomatic weapons as permitted by the state laws where the incidents occurred.

There is a third issue: America's gun culture and the glorification of these weapons, which makes law-abiding citizens want to own them. Even I have entertained the idea of buying these weapons. And those law-abiding citizens make this issue much more complicated and less about a simple idea that assault rifles are bad, a perspective taken by many liberals. Of those 18 million "modern sporting rifles" in society, the vast majority are in the hands of law-abiding citizens. That is why efforts at gun control legislation fail when it includes bans or discussion of buy-backs of semiautomatic rifles.

Remember the question about data from the first chapter? How often do legal gun owners use guns to defend themselves? The numbers of confirmed reported cases vary from 50,000 to 70,000 a year with about 1,500 to 2,000 ending with someone dying. No one tracks these events with the precision that the FBI tracks the national crime statistics, so the numbers are a little vague and they differ depending on which source you are using. There are even claims that the number is much higher due to the number of times in which "the victim pulled a gun, and the assailant ran away and no police report was filed." This is a gray area of opinion surveys and some researchers have even put the number of defensive gun uses in the millions per year. The problem is that it happens often enough where gun owners firmly believe that it is a valid position and there are multiple verified instances where gun owners do defend themselves. But there are also multiple instances where the criminal uses the victim's own gun against them. Again, you can find anecdotes to support almost every side of a policy question, and without hard accurate data, both sides feel they are correct.

The Supreme Court has ruled that the Second Amendment doesn't prevent a local jurisdiction or the federal government from establishing laws limiting the type of guns or requiring that citizens shall secure them or register them. There are existing laws that require sellers of guns to be registered, and registration for machine guns doesn't fall into the category of personal weapons. You can own a machine gun or even a cannon; you just have to go through a federal registration and background check that is supposed to be very thorough. The fact is that we regulate and track automobiles better than we do guns. Every car on the road must be licensed, registered, and insured. Any private sales between individuals must be reported to transfer the title. The number of unregistered cars is less than 5%, and some of these are race cars, since they don't need to be registered for racing on private racetracks (hence the term "street legal" for a race car that is registered). The number of unregistered cars on the highway varies but is thought to be less than 3%. That would be about 7.5 million of America's 250 million vehicles. Having been driven around in an unregistered "ranch truck" on a friend's property in Texas, I believe that number and most likely they don't ever touch paved roads. But compare that with the little over 6.6

million registered guns of the 400 million that have been sold in America over the years—that's 98.3% unregistered. And when we say *registered*, we mean they were voluntarily registered; very few states require guns to be registered.

Mass shootings are a very emotional issue, especially when they take place at a school. One of the problems is that the definition of *mass shooting* changes depending on who you talk to. The Congressional Research Service (CRS, Policy Research for Congress) defines it as four or more victims shot and killed in one incident, excluding the perpetrators at a public place. CRS excludes gang-related killings, acts carried out that were inspired by criminal profit, and terrorism. Using that definition, much of what most people consider mass killings gets excluded.

At the other end of the spectrum is the Mass Shooting Tracker (MST), which defines a mass shooting as any event in which four or more people are shot, including the perpetrator. When you hear very high numbers for mass shootings, you are likely getting it from MST.

In 2019, CRS said there had been eight mass shootings with sixty-nine deaths and ninety-three injured. Contrast that with MST indicating 503 shootings, 629 deaths, and 1,901 injuries. The problem is both sets of numbers are too far to the extreme. CRS excludes killings in home, and there have been a lot of the those, while MST lumps in events where people were only wounded or were really tied to criminal activity. We are going to use three people killed excluding the perpetrator and include domestic, terrorism, events over multiple days, multiple shooters, and unsolved cases. Based on that, 2019 had fourteen events, ninety killings, and 126 injuries. And we are going to look all the way back to 1966 to include the University of Texas shooting.

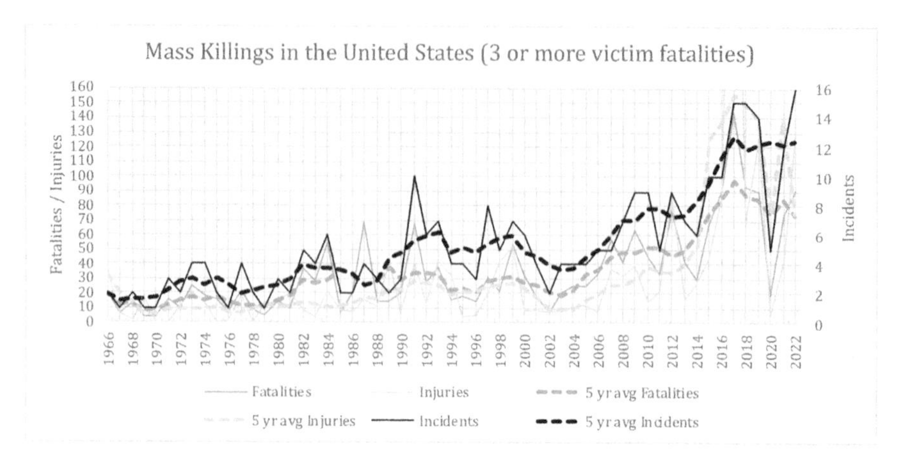

FIGURE 1. MASS SHOOTINGS

After every event, people will howl that it is the guns, and we must do something about the guns. But this chart shows us two interesting points, a drastic drop in events in 2002 and 2020. We will look at 2002 first, specifically after September 11, 2001. Because after 9/11, there were no mass shootings for the rest of the year and only two the following year. The volume of guns didn't change, so what did? Was it all the security? Well, security was beefed up at airports, but it wasn't in schools, churches, or places of employment. My guess at what changed is the level of hate. The country came together, partisan attitudes were set aside, and the country worked as one.

A mass shooter needs three things: the weapon, a target with the potential for lots of casualties, and the hate in his heart to do the deed. There is no doubt that lax gun laws in many areas have increased the volume of weapons available. There are multiple examples of people buying guns and using them in a mass shooting event within days, sometimes hours. All too often those are high-velocity semiautomatic weapons with large-capacity magazines, so we do need to tighten the controls on guns, but that alone won't do it. We also need to consider the hate. The growth of social media and online forums where hate can fester has tracked with the increase in mass shootings. There is no getting around that the hate being pumped into our social space by politicians, pundits, and media groups is being amplified by social media and has had an impact. Think that's wrong?

Well, what about 2020? The hate levels were the same, the guns were still around, so what happened? With COVID hitting the country and a vicious partisan presidential election on the horizon, we saw a drop in mass shootings. But why? The guns were still present, but for most of the year, public places were closed. Workplaces and residences were four of the five settings for the mass shooting events that did happen, but why weren't there more? There were, though they didn't fall under our definition of mass shooting, but they did meet MST's definition. MST reported 696 events, with 661 fatalities, and 2,750 injuries for 2020, which was higher than the 2019 numbers and lower than the 2021 numbers. The hate was motivating people to pull the trigger; it was just that the groups of people were much smaller.

You can already see some people moving to harden or reduce the targets that could generate multiple fatalities. The idea of increasing security against mass shootings is coming from the conservatives to avoid implementing gun control. Little wonder, the NRA's meetings and conventions do not allow guns inside. The same is true of the Conservative Political Action Committee (CPAC) when they hold their major gatherings. More security on schools, churches, and public events are being pushed as a solution, but they may have an unintended consequence. A big push is to make schools have a single entrance with security. While that would avoid the few mass shootings, it would make schools much more limited for fire evacuations. And let's be real. Between 2014 and 2018 the National Fire Protection Association reported an average of over 3,200 school fires per year with on average one casualty per year and almost forty injuries.[28] Making schools more difficult to evacuate will increase the risk of fires to solve the super-rare school shooting. This would be the height of folly.

Other chapters will look at social media and discuss what can be done to lower the hate in public spaces. As far as guns, as badly as people want to confiscate "assault weapons," that isn't going to happen. As we said earlier, you are punishing the country for the actions of the few. It would be the embodiment of the fear that the conservatives and libertarians have been warning about and would not only violate the Second Amendment, but it would also split the country. But we do have some suggestions at the end of this chapter.

The last issue to consider is children's access to guns. There has been a push to require gun owners to secure their weapons using trigger locks or lock them in a secure cabinet. There are numerous examples of children accessing unsecured guns and shooting themselves, others, or taking the guns to school. In some countries, strict laws require that guns be secured, and that ammunition be secured separately from them. Most gun courses stress this as a safety measure. The reality is that too many people carry loaded guns or store them in their cars or bedrooms with an idea of defending themselves, and they believe that the idea of locking the gun and ammo separately defeats the purpose.

So, what is to be done? It comes down to managing risks and consequences. It also comes down to a set of laws that would not apply equally to everyone. If a law was passed requiring guns to be secured, based on children, then it really wouldn't apply to childless adults or those with grown children. The fact is that about 600 children die each year due to suicide or accidental shootings, and these are horrific events. But are they sufficient to pass a law requiring all gun owners to secure their guns or suffer legal action if they don't?

For the family of a child who has just committed suicide or been accidentally shot, it may seem insane to charge the owner of the gun with a crime, but that is the answer. Especially when you consider that in many cases, the gun owner is not the parent of the victim but was responsible for the shooter getting the gun. The solution is a simple law saying if your gun was used in a shooting, then you are responsible for failing to secure it or neglecting to report it as sold or stolen. It would function in the same way as vehicular responsibility. Let someone borrow your car, you and your insurance is on the hook for any accidents caused by that car, especially when it is your underaged son driving.

If we want to get accidental shootings and suicides by children under control, this is the answer, not a law that requires securing guns and could only be truly enforced by searching everyone's houses.

Gun control is like so many other issues where the false claims aren't limited to one side. Second Amendment advocates for reducing gun laws point out that Chicago and Washington DC's strict gun laws haven't prevented gun homicides, but that doesn't recognize that the nearby regions tend to have very loose gun laws. And as we discussed earlier, criminal gun homicides are more about poverty than anything else and mass shootings are more about a single disgruntled or disturbed individual's ability to cause large numbers of casualties.

This brings up the question of what to do about automatic weapons and weapons in the hands of unstable individuals. There has been a push for "red flag" laws where individuals can have their guns taken away if they are identified to be a threat to themselves or the public. The problem has been that these laws have been enacted in some states and to varying degrees and usually aren't clear on when the decision to remove the guns is made and how a person applies to have their guns returned.

The Justice Department should draft guidelines, have them reviewed by the Supreme Court, and then send them to the states. It would then be each state's responsibility to either adopt, modify, or reject the guidelines. That may seem like a weak answer, but realize that this is a jurisdiction of the states. We cannot look to the federal government when the real solution is to resolve it at the state level. That does have the possibility of fifty different variations, but it respects the actual division of powers that lies at the heart of our Constitution.

Here is the summary of the data and logic so far:

- The United States has over 400 million guns in private hands.
- The number of guns owned doesn't relate to gun homicides.
- The number of guns owned does relate to gun suicides, but not to the overall number of suicides.
- There are more than 18 million semiautomatic legal versions of fully automatic military rifles in private hands.
- Mass murders constitute a very small (615 in 2020) part of gun homicides, but those that occur at schools and involve innocents have a large social impact.
- Most guns in America are unregistered.

It is time for liberals to give up the idea of forcing the registration or confiscation of America's guns. It is simply not feasible. First, there is the cost of taking millions of guns off the street. Guns can range in cost from hundreds to thousands of dollars. Plan to repay the owners for the cost of their weapons? You could be looking at over a trillion dollars. Then add the expense of all the manpower to remove those 400 million guns from, say, 150 million owners. Second, the legal costs: the Second Amendment is part of the Constitution, and the Supreme Court has upheld it repeatedly. It is impossible to think that a constitutional amendment could be approved and even more unlikely that any court would allow a massive gun confiscation. Now add the political and emotional costs: this would deeply divide this country. That is why liberals need to stop demanding bans on gun sales and any form of confiscation.

But it is also time for conservatives to stop blocking every gun control effort and framing it as if it was some type of liberal witch hunt. As Ronald Reagan said, "With the right to bear arms comes a great responsibility to use caution and common sense on handgun purchases."

The first necessity would be to stop the supply of high-capacity magazines for both pistols and rifles. Like bump stocks, they make the potential for mass murder too easy. How are a hundred rounds necessary for hunting, target-shooting, or anything legal? If you feel the need for a hundred-round drum magazine to go hunting, it might be time for a marksmanship course. Large magazines for semiautomatic pistols or rifles have been at the heart of too many of the most horrific mass shootings, and stopping new sales is the least that the gun community can do. Liberals may feel the need to ban new semiautomatic weapons, but there are already 18 million in civilian hands, most belonging to law-abiding citizens. Banning semiautomatic weapons won't change anything, and it goes back to the very basis of the Second Amendment. Instead, leaders should address poverty, hatred, and the mental health crisis—the root causes of gun violence.

The second requirement would be background checks and cooling off periods. Someone can come in and ask to buy a gun and register, but then allow time for both the background check and the purchaser to consider whether they need that gun. The NRA, gun manufacturers, and gun sellers will object to this, but it is needed. There are just too many cases in which someone purchased a weapon and then immediately used it to shoot someone. A two-to-four-day period gives the background check time to be done properly and confirmed, and it gives the purchaser time to determine whether they can really afford the gun and if they really need it. Impulse buying and firearms is not the best combination. This requirement would extend to background checks for gun sales between individuals and at gun shows, even using a system where people can pre-register or with shorter wait times.

The third thing to consider would be that all states ban the open carry of semiautomatic rifles. With the rise of private militias and radicals supporting conspiracy theories, how is it a good idea that these people could walk up to a state capitol or crowded location with an openly displayed semiautomatic rifle? That is something that is going to give state troopers sleepless nights after the political rallies and the incident at the Capitol. If a state allows the open carry of pistols, shotguns, or bolt-action rifles, they should remain legal to carry openly. But it should not be allowed to openly carry semiautomatic rifles that could be converted to full auto. There have been too many incidents where police said they were outgunned by criminals. The potential for a group of radicals to legally bring that level of firepower to a political demonstration or any mass gathering is too much of a risk, and it disadvantages law enforcement.

The fourth thing is for liberals to stop establishing "gun-free" zones around schools, churches, or other gathering places. In 1990, Congress passed the Gun-Free School Zones Act, which was ruled unconstitutional by the Supreme Court in 1995. The law was revised and repassed in 1996. It prevents unlicensed individuals from having guns with 1,000 feet of kindergarten through twelfth-grade schools. It allows authorized people (law enforcement officers) and people who have been licensed (concealed carry permits) and it only applies to unlicensed possession. The issue has not prevented the school shootings that have occurred since 1996 and will not prevent extremists from attempting an attack again. As written, the law impacts law-abiding citizens not only on school grounds, but within 1,000 feet on public sidewalks, roads, or highways. Most states allow some form of open carry, concealed within a vehicle or without a permit, this creates a legal problem. A "gun-free" zone is almost unenforceable for police officers; they can't stop and frisk everyone, they can't check every vehicle as parents pull up to pick up their children. So, the reality is it makes people driving by on nearby streets violators of a law when they may not even be aware of the school on the next street. It is an unenforceable and unnecessary law which provides a false sense of security and fails to address any real problems. Instead of spending money to put up signs and calling areas "gun-free," we need to address the underlying problems of poverty and mental health.

The fifth necessity would be to establish a national concealed carry permit, not to override jurisdictions that don't allow it, but to provide uniform training and requirements. Each jurisdiction can then decide to recognize the national permit, have their own with different requirements, or ban concealed carry altogether. And each jurisdiction would register their decision with the national agency so anyone could look at the national agency's website and know before they entered whether it was banned, local-permit only, or if the national permit would be accepted.

The sixth necessity would be to have the Department of Justice draft a law that allows legal action against individuals who fail to secure their guns or to report the theft, transfer, or sale of their guns and those guns are subsequently used in a crime or accidental shooting by others. The Department of Justice should work with states that have already enacted such laws to determine what has worked, what hasn't, and what needs to be improved to make the laws effective while also respecting individuals' rights, responsibilities, and the Second Amendment. The law would be reviewed by the Supreme Court and sent to the states for each state to adopt if chosen.

The seventh necessity would be to have the Department of Justice draft a guideline for red flag laws that would define when guns could be removed from an owner, how removal would be initiated, and which local government entities would have jurisdiction to decide on removal. Again, the Department of Justice should work with states with existing "red flag" laws to learn to define the guidelines. It would also define recommended procedure on how the gun owner would recover his guns from the local law enforcement agency. The guideline would be reviewed by the Supreme Court and sent to the states for each to consider and adopt if chosen.

A final note: A friend told me about a concealed handgun class he went to. It began with the instructor handing them a slip of paper and having them write down all their assets—house, cars, boats, jewelry, retirement accounts, etc. Then he made them roll up the paper and put it inside the barrel of their guns. He explained that is what's at risk every time you draw the weapon and fire at somebody. After that, my friend paid more attention to that class than any other class before. That needs to be part of the national concealed carry permit.

13.
LGBTQ+

It is absolutely imperative that every human being's freedom and human rights are respected, all over the world.

—Jóhanna Sigurðardóttir

The last of the hot button issues is the ongoing culture war around lesbian, gay, bisexual, queer, and transgender people. The term LGB began in the 1980s and has been expanded as more groups were folded into the overall community, which has now adopted the alphabet soup of LGBTQ+ and continues to expand.

The history of homosexuality has been of oppression with some short periods of tolerance. The Code of Assura, laws from the Assyrian empire of 1,500 to 1,100 BCE, punished homosexual acts or even falsely accusing someone of one. In the Bible, the Book of Leviticus states that "a man shall not lay with man as with a woman." That one reference is the basis for the Christian views against same-sex relationships that have existed in Western society to this day.

There was a rare period in ancient Greek society (700 BCE to 200 BCE) that openly accepted homosexual acts. That was when there were a few military units that were completely gay. When he was conquering Greece in 328 BCE, Phillip II of Macedon regretted that his army had wiped out the Sacred Band of Thebes, 150 pairs of gay lovers who had been an undefeated elite unit of the Theban army. When the Roman empire conquered the region beginning in 146 BCE, this marked a return to official repression of gay and lesbian activity as Roman law made same-sex relationships illegal.

Outside of this short era in Greece, being gay or lesbian was sometimes tolerated discreetly, but almost always officially illegal. Consider the case of Oscar Wilde. He was tolerated in Victorian society until he became embroiled in a libel suit with the Marquess of Queensbury over whether he was gay. Once his homosexuality was exposed, he was sentenced to two years of hard labor in an English prison from 1895 to 1897. After being released, he left England and never returned.

In America, a similar attitude existed until the 1960s and 1970s. Gays and lesbians were tolerated as long as they kept their sexual activity hidden. Sodomy laws and a serious prejudice against gay men was prevalent. In 1960s and 1970s, homosexuals became more open and participated in society openly. Unfortunately, laws still existed banning same-sex relationships in most states, and in 1986, the Supreme Court upheld a Georgia law that made sodomy illegal.

Even as the Supreme Court was handing down this decision, society was changing. The country was becoming more accepting of same-sex behavior. It didn't happen overnight, but by 2003, the Supreme Court struck down Texas's sodomy laws, and by extension reversed its 1986 decision and invalidated the remaining sodomy laws in thirteen other states.

While mainstream society has become more accepting of gays and lesbians, there are still fears and hatred of the gay/lesbian/bisexual community in some areas. The Pew Research Center conducted a global study on acceptance of homosexuality in 2019. Results showed that 72% of Americans said homosexuality should be accepted by society. Argentina (76%), Canada (85%), Australia (81%), and a number of Western European countries (though not all) scored higher than the United States, but outside of them and the Philippines (73%), the rest of the world scored worse, far worse in many countries.[29] And while acceptance has improved in the United States, this has become more complicated with growing conservative angst especially about transgender people and the concerns over the ability to medically change a person's sex.

Demographics are that the lesbian/gay/bisexual community is likely 4.5% of the total population and transgender people are about 0.6% of the population. The Wikipedia page on LGBT demographics of the United States quotes a 2017 Gallup poll for these numbers. The reason we don't have specific data is the 2020 US Census had planned to ask about sexual orientation and gender identity but scrapped those plans in March 2017. While some have claimed that this was at the request of the Trump administration, no confirming evidence has been found. As a result, we only have polls and guesswork to determine the demographics of the LGBTQ+ community, rather than hard census data.

This part of the culture war breaks down into some subtopics:

- the legality of denying services to LGBTQ+ community based on religious objections,
- the use of pronouns,
- transgender people in sports,
- the ability of minors (under legal age) to use gender-changing drugs or operations.

The first issue is the ability of people who object to LGBTQ+ lifestyles to withhold services. There have been two public cases where couples planning same-sex marriages were turned down for a wedding cake in one case and flower arrangements in the other. A third case was brought by the conservative legal group Alliance Defending Freedom and challenged the Colorado law at the heart of the wedding cake case. All the cases went to the Supreme Court. In 2018, in the cake case (*Masterpiece Cakeshop v. Colorado*), the justices ruled seven to two that the baker could not be forced to make a cake for a couple when his religious views indicated that same-sex marriage was immoral. However, no guidelines were provided for lower courts to follow.

The 2021 case was *Ingersoll & Freed v. Arlene's Flowers, Inc.* When it was first brought to the Supreme Court, the justices ordered the Washington State Supreme Court to review the case based on the Masterpiece Cakeshop ruling. The Washington State Supreme Court upheld its decision against the florist requiring her to provide services to same-sex couples. In July 2021, the Supreme Court decided not to hear the appeal from the florist, thereby leaving the Washington State Supreme Court decision in place. Which leaves us with the paradox: you can't require a baker to bake a cake, but you can require a florist to make flower arrangements for same-sex marriages. Does this seem confusing? But wait, the Trump administration issued Health and Human Services rules in 2019 that allowed personnel and institutions to refuse services when their conscience was being violated in order to protect their religious liberty. This was considered outrageous by many because it was viewed as allowing discrimination and could result in some medical providers refusing care. It seems unacceptable to allow a fellow American to die just because you don't agree with their lifestyle or views.

So, we end up with two ends of the spectrum. The first is people providing critical services: medical care, emergency services, and requirements to live. Can they refuse service? Most Americans would consider it completely unacceptable to deny necessary life-sustaining requirements.

But the other end of the spectrum is the artist. I don't think any of us would think it is reasonable to require an artist to paint something that he found offensive if it had been requested by a patron. Art is a personal expression, and requiring an artist to create art that objects to their personal views seems totalitarian.

So, where is the crossover point? Can a grocer refuse to sell groceries to someone? Can a plumber refuse to fix someone's house? And what are the individual rights of a service provider and the person requesting the services?

There are some examples to consider. Several businesses clearly indicate "no shoes, no shirts, no service." But do we want to take this to the point of a business declaring that they have issues of "conscience" in serving specific people? If the florist had said on her front door and website that same-sex marriages were excluded, would that have made it acceptable? A complicating factor is that one spouse of the same-sex couple was a longtime client of hers. So, it was okay to sell him flowers, but not to do his wedding? And if the florist had indicated that beforehand, would that have resolved the issue?

The problem is further complicated by allowing people to deny service by religious objection. How far does that extend? Would we allow a business to exclude someone based on their race? What about excluding them because they follow another religion that the business objected to? Would that be fair? Would it be just?

Welcome to the minefield between the two ends of the spectrum. In the 2018 Masterpiece Cakeshop case, Justice Kennedy, who had written every major Supreme Court decision protecting homosexuals, wrote that the Colorado Civil Rights Commission had acted inconsistently in cases involving opponents of same-sex marriage, "concluding on at least three occasions that bakers acted lawfully in declining to create cakes with decorations that demeaned gay persons or gay marriages."[30] Think those three cakes were setups to reveal a bias? Very likely, but the Colorado baker is back in court. In June 2021, a Colorado court ruled that when he refused to make a birthday cake for a transgender woman, he violated the law and was fined $500. Think this was a setup by LGBTQ+ activists to provoke another court case? Both extremes in this part of the culture war are employing a scorched-earth policy in the courts and social media.

Which brings up the third case brought by Alliance Defending Freedom. The issue here was a website designer who wanted to refuse to make websites for same-sex weddings, despite that same Colorado law that bars discrimination based on the sexual orientation, race, gender, and other characteristics. The Supreme Court indicated that the act of designing the website was speech and Colorado was requiring the web designer to perform that speech for same-sex weddings, which ran contrary to her views that marriage was only between a man and a woman. The designer had even stated that she "provides website graphic services to customers regardless of their race, creed, sex, or sexual orientation." Thereby focusing this down to the issue of same-sex weddings and referencing the Masterpiece

Cakeshop law to show that Colorado has tried to force business to "compel this speech in order to excise certain ideas or viewpoints from the public dialogue." The six conservative justices on the Supreme Court ruled that the Colorado law was unconstitutional in requiring the web designer to provide marriage websites for same-sex weddings. Justice Gorsuch wrote in the majority opinion:

> In the past, other States in Barnette, Hurley, and Dale have similarly tested the First Amendment's boundaries by seeking to compel speech they thought vital at the time. But, as this Court has long held, the opportunity to think for ourselves and to express those thoughts freely is among our most cherished liberties and part of what keeps our Republic strong.[31]

So, what is the answer? Maybe it is that businesses need to post both at the front door and online if there are groups who they will not serve and accept the repercussions. The Colorado baker indicated that he had lost business and had to lay off employees due to the original cake case that started in 2012. Did he lose business due to the legal case or because his customers learned he had religious objections to the LGBTQ+ community?

The first step is to designate the essential services. Just as during the pandemic, it was necessary to decide on a minimum requirement for society to function. Everyone employed in essential services should recognize that they can't discriminate based on anything. That is the whole point of essential services—they are essential and cannot be withheld from anyone without threatening their ability to live.

Everything else is nonessential, and if a business chooses not to serve someone because of their sexual preference, manner of dress (no shirts, no shoes), or any other reason, they need to clearly post it, both in person and online and then accept the consequences.

Some are going to argue that this isn't acceptable and that a business needs to be open to all if they are open to the public. And for essential businesses, that would be correct, but this misses the point: we do allow nonessential businesses to limit their clientele based on several factors. It's hard to allow a business to publicly set out who it will and won't serve when that could extend into areas that many would consider discrimination, but it adheres to the principle of personal choice and responsibility. While the counter-argument as stated by one person—"We all have the same right to the same cake. . . ."— is correct, but do you have the right to force the baker to make the cake? This is the true question here. Especially when there are other bakers who would be more than willing to bake the cake for that client.

The fact is there are people with clear biases, whether they are based on how they were raised, their religion, race, ethnic views, or anything else. To force them to serve the people they have a bias against may seem right to some, but what is it really doing? Is it helping them overcome their biases or just forcing it in their faces and angering them more? The question then boils down to this: Do we want people with suppressed, covert biases, or is it better to just let them come out in the open and admit that they are biased? The baker, the florist, and anyone else can make a choice not to serve certain people. Clearly state it and post it, and then accept the consequences. No one really needs to protest in front of their shop. No one needs to mount an online campaign to put them out of business. We don't need more expressions of outrage in this country in response to their choice. We just need to trust that the increasing acceptance of the LGBTQ+ community and the changing demographics in this country will slowly address this issue without the need for more hate or anger.

I saw that when my Midwestern parents moved the family to New Orleans and after a few years, opened a small printing company. This brought them in contact with several homosexuals who ran charities and businesses in New Orleans. My parents had always been fiscally conservative/socially liberal Republicans and believed that what two consenting adults did in their bedroom was none of anyone else's business. But working and interacting with the New Orleans gay and lesbian community did change them, and I personally witnessed it. When some of our Midwest family were visiting New Orleans, a relative made disparaging comments about homosexuals. My mother defended some of her best customers in a quiet but steely voice, shocking the relative and the entire family into silence.

This idea of letting biases come out in the open may offend many, but let businesses post who they will and won't serve. Most people will prefer to have expressed biases out in the open, rather than allowing them to continue to be expressed indirectly from the shadows.

Some members of the LGBTQ+ community are asking to be called by specific pronouns to express how they view themselves. And sometimes these pronouns differ from the gender they were born as. This contrasts with cisgender identity, in which a person's view of their gender matches what they were born as. Over 99% of America is cisgender and for some of the older and more conservative Americans, it can be disturbing to encounter people using pronouns that differ from tradition. That is where this whole backlash against transgender and gender-fluid people originates. Remember that the older generations have experienced massive upheaval in their worldviews and just accepting cisgender homosexuals has been a big shift for them. To then call someone *she* when they were born male or to call them a gender-neutral term when they were taught to call all men *sir* may not seem like much, but it is asking for more change. And in some cases, the LGBTQ+ community isn't asking; they are demanding change. And the increased vitriol and anti-trans local legislation from conservatives and the counter-response from progressives is escalating the issue.

Forcing people to change has not worked well in the past. While it is perfectly fine to indicate that someone prefers to be called a certain way, requiring that others use that form of address is where lines are starting to get crossed. We are trying to reach into someone else's mind and unwind years of learning, habits, and behaviors just so they change how they call someone. Most men raised in the '60s and '70s, especially in the South, were taught to call people that appeared to be men *sir* and women *ma'am*. It's a habit that I still have to this day. It's something that is done instinctively without thought or prejudice.

Everyone should have the ability to request they be addressed with their preferred pronouns, but they need to accept that not everyone will do it. And rather than view it as a personal affront, they should view it as the other person's inability to deal with change.

Which brings in the next issue. The anti-trans legislation being pushed by conservatives breaks down into three main areas: youth sports, bathroom access, and changing the gender of youths.

The core issue with youth sports is children born male who identify as female and compete against females when their inherent maleness gives them an unfair advantage. That may seem one-sided, but you never see concern over people born female competing with the boys. As we said earlier, trans people are less than 1% of the population, so these cases are incredibly rare. So, why are state legislatures even attempting to address this? Because the issue resonates with core conservatives who feel threatened. In the last forty years, court rulings have struck down sodomy laws, upheld same-sex marriage, and found several laws to be discriminating, setbacks to highly conservative voters who feel that the way of life they grew up with is slipping away—especially when you consider that most of these voters are in rural areas where trans people are non-existent. So, if there is no threat of a trans person running on the local girls' track team, why are we having this discussion, and why are progressives waging war against these laws in states where the legislature is controlled by Republicans? Because it is a culture war with no prisoners, no compromises, only victims.

So, stepping back from the culture war, where does the science, the history, and data take us?

For international athletics, in the 1950s, sex verification testing was adopted in the form of physical examinations. But in 1968, due to protests over the intrusive nature of physical examinations, the International Olympics Committee (IOC) changed to chromosome testing, because women have two X chromosomes, while males have an X chromosome and a Y chromosome. Sounds simple, right? Not totally. The tests were found to be inaccurate and inconclusive. This has now evolved into testing for the level of testosterone. There are two problems with this. The first is that all fetuses start out undifferentiated (neither male nor female) and the SRY gene, which is part of all human DNA, determines a person's gender in utero. But for some people, less than 1% of the population, the SRY gene doesn't act or is mislocated on the chromosomes. The result is a long list of conditions possible for that >1%, one of which is hyperandrogenism where females have higher levels of testosterone. This >1% of the population is referred to as *intersex*, individuals whose gender at birth falls between the classic definitions of male and female. And this is not a new condition; references to hermaphrodites and androgynous individuals reach far back into history.

To complicate things, when the IOC did not test testosterone levels during the 2016 Summer Olympics, all three medalists for the women's 800m race were intersex. Studies of athletes have found that for males, testosterone levels vary greatly, but on the lower end, they actually overlap with some women athletes who show elevated levels of testosterone. At what point does an athlete competing in women's sports gain an unfair advantage over other women? Rather than banning all trans athletes, maybe it should be up to the medical and sports professionals to answer that question. And let's keep that question out of the hands of state or federal legislatures.

Same thing with bathroom bills that try to forbid trans people from using a bathroom that doesn't correspond with their birth gender. Think about it, most trans individuals appear to be the gender they identify with. How would a bathroom bill be enforced? The idea that this is somehow protecting women in the sanctity of the ladies' room misses the point. We have existing laws against rape and violence, but how would a law about preventing people, some of whom look like women, from entering the ladies' room work? It isn't enforceable. You can't have a police officer outside the door 24/7 checking people as they enter. It only provides a false sense of security to cisgender people and works to embarrass and harass a trans person who needs to use the restroom. Which is why it has been rejected or overturned in all the states that tried to implement it. State legislatures are trying to control the gender of an amazingly small part of the population.

The final issue is the use of gender-altering drugs or surgeries on people under the legal age defining an adult. Some conservative states have been moving to limit or block when people can get gender-changing drugs or surgeries. A few states are moving the opposite way to lower age restrictions to allow younger teens to receive what are referred to as gender-affirming surgeries or drugs. There are two sides to this argument, and each is quick to bring out stories emphasizing either when these surgeries work well or when people regretted them. We need to go back to the science and data. One article pulled together data from twenty-seven studies, looking at almost 8,000 patients who underwent gender-affirmation surgeries and found that only seventy-seven regretted it, fewer than 1%.[32] The authors are quick to praise that this low regret rate is due to the selection criteria for these surgeries, while those opposed to these surgeries are just as quick to dramatically recount the stories of those with regret.

The medical community can debate the degree to which the fewer than 1% affected by the SRY gene overlaps with the fewer than 1% of trans individuals in America, but there is an overlap between these two groups. While conservatives may be offended by the idea of changing a person's sex, there is a small portion of the country whose genetics have left them in the gray zone between male and female. And for these intersex individuals, gender-affirming surgeries and drugs may provide an opportunity to lead a better life. The issue is at what age someone should be able to have that surgery, and again, this is something best left to the individual in conjunction with their doctors and family. The reality is that people mature at different rates, and looking at the issues of what age to allow people to drink alcohol and vote should show the difficulty in establishing a minimum age for all youths.

So how to resolve the issues? If there is a middle ground, then it has to encompass two basic principles. First, that conservatives need to accept that the LGBTQ+ community is not going back into the closet and to recognize that like every other minority, they just want to be treated fairly.

Second, progressives need to recognize that you can't force someone's views to change and backing them into a corner and threatening them with jail or financial ruin is only going to make them dig in deeper and fight harder. The hypocrisy of the progressives is that while they champion the LGBTQ+ community in coming out of the closet, they seem perfectly ready to push the anti-LGBTQ+ community into a closet of suppression. The resulting conservative backlash is perfectly predicated on this reaction. The question is, do you want covert biases or overt biases? My view is that it is always better to get things out into the open where they can be clearly seen.

The policy for this section is as follows:

- Businesses cannot discriminate against anyone if they are designated as essential. Nonessential businesses shall clearly post both at the business and online their rules of who they serve and who they will not serve. A nonessential business shall not turn people away unless they have clearly posted their rules of service, both physically and online.

- Federal and state legislatures have no business legislating the definition of male/female for intersex/trans individuals in regard to youth sports, access to bathrooms, or legal age for gender-affirming surgeries/drugs without a clear scientific consensus from the medical and trans community.

That completes our journey through the culture war issues. Throw in Critical Race Theory (CRT) in education, Diversity Equity and Inclusion (DEI) in college admissions, Environmental Social and Governance (ESG) in investing, and whatever new issue du jour is driving our culture warriors and you have pretty much covered the items being used to separate us into warring tribes. And as we have shown, these items tend to arise from much deeper issues of poverty, overpopulation, and concerns over changing demographics, problems that these surface issues avoid and do not solve. And as we will see in subsequent chapters, those deeper problems are the ones that are really causing us issues and if we can address the deeper problems, it would cause the culture war issues to fade.

Section III: Social Policies

Society: (noun) the community of people living in a particular country or region and having shared customs, laws, and organizations.

—Oxford Languages

The central conservative truth is that it is culture, not politics, that determines the success of a society. The central liberal truth is that politics can change a culture and save it from itself.

—Daniel Patrick Moynihan

14.
College Education

One of the greatest disservices you can do a man is to lend him money that he can't pay back.

—Jesse H. Jones

Education must be a priority. Liberals have been advocating for free college education and the forgiveness of student loans. It has gained a lot of support, but there are certain facts that need to be considered.

In fall 2020, there were approximately 19.7 million college students in the United States with about 14.6 million enrolled in public colleges and 5.1 million enrolled in private colleges. These numbers are about the same as in 2018, so enrollment numbers are holding constant. The average tuition and fees for a public college education is about $10,000 per year. We will assume that liberals don't plan to pay for private school as well, so we're looking at the 14.6 million students at public institutions. Basic math gives you a spending of $146 billion per year. With free education in public institutions, we could easily see an increase, possibly a massive increase, in public university enrollment, exceeding $150 billion per year. This is not a one-time cost; it is an annual expenditure that cannot be supported under the present budget. So where do you get $150 billion without increasing the deficit?

But wait, that's just tuition and fees, it doesn't include meals, accommodations, books, or any of the other costs of college life. For a public college, additional costs for room and board can add another $10,000 per year. And remember that most of those students are getting college degrees to achieve high-paying jobs. Of the students enrolled in four-year institutions, over half are in fields that would result in high-paying jobs in engineering, healthcare, law, or business.

Another consideration is that of those 14.6 million students, many do not complete their degrees. The National Center for Education Statistics estimates that the overall graduation rate is 62%. This is based on the number of students who achieve a bachelor's degree in six years or less. So, over a third of students aren't finishing college. For part-time students, numbers are worse. Within an eight-year time frame, only 23% achieved college degrees. Those number have been improving very slightly. The concern is that if college is free, then more students will attend, overwhelming schools and making it harder for those who are actually capable of completing college and achieving a degree. People will point to several examples of free college around the world, but most are in Europe and not all of them pay the non-tuition costs, which means that students still need loans, jobs, or family support to attend college. Fewer students go to college in Europe, which has a larger combined population than the United States, and some countries have better developed trade schools to train technicians and workers for jobs that do not require a college degree. Free college in America would be a very different thing and could end up with a lot of unqualified students (both rich and poor) clogging up the system, thereby detracting from the experience of qualified students, and worse for the government driving up those college costs.

Some experts actually blame the easy availability of federal loans for the existing increase in costs, indicating that this has meant there is no incentive for colleges to constrain or reduce costs to maintain enrollments. A shift to "free" college where the federal government is picking up the bill wouldn't provide an incentive to control costs either, which could spiral into more federal involvement with local colleges—not less.

That brings us to student debt. As of 2020, 42.3 million Americans owe a combined $1.5 trillion of federal student debt. The volume of debt has almost tripled from the $590 billion owed in 2007. One reporter pointed out that wages have been barely increasing when adjusted for inflation, while college education costs have been increasing at a much higher rate.[33] The cost for a four-year degree including room and board in 1989 was approximately $26,902 (about $52,892 in modern dollars). Compare that to over $100,000 for the same four-year education in 2022. During the same period, once adjusted for inflation, wage growth was only 0.3% per year. "Many Baby Boomers and Gen Xers," he writes, "remember working their way through college and graduating with little to no debt. Sadly, that feat is virtually impossible for the current crop of students and recent graduates." The debt crisis is critical in private, for-profit institutions where 93.1% of the present 992,000 undergraduates are receiving federal aid. The institutions with the highest debt among their student bodies are all private for-profit. The existing federal loan system for college education is broken, and it hasn't served the student body or society well. With so many in debt and approximately 10% of them defaulting on the loans, most people agree. So what is the answer?

The first question is what is the objective? Is it to send everyone to college? With dropout rates at their current rate, that doesn't seem to be feasible currently. The dropout rate can be attributed to public secondary schools not sufficiently preparing students for college. The need for doctors, lawyers, engineers, and all the other required specialists to help society function is high. Using federal dollars to get a degree in sports management with an eye for becoming a sports agent doesn't seem like something that society should pay for, unless I have missed a massive shortage of sports agents impacting the economy.

The use of federal student loans in private for-profit schools simply needs to stop. There is too much abuse, too many issues with how the schools operate, and the cycle of debt and low graduation rates indicates that it needs to stop. The question is what to do for private non-profit and public colleges. The fact that federal student loans may have fueled an explosive rise in college costs means that change is required, and the first step is to stop issuing federal student loans and scholarships as they are presently done. It wouldn't be an immediate stoppage; it needs to allow students now in college to complete their degrees as the new system is set up. So, what are the consequences of that action?

Upper-income and upper-middle-income families take advantage of federal student loans, but they don't absolutely need them. Private financing is available for those pursuing high-paying jobs who have someone to co-sign (i.e., family backing). Private non-profit schools already offer scholarships and should be able to offer loans. But something is needed to encourage and enable low-income and middle-income families to send their deserving and qualified children to public college without bankrupting the family or government. *Deserving and qualified* means able to enter college and leave with a degree. College is not for everyone. Having personally searched high and low for skilled welders and pipefitters for large industrial projects, I can attest to the fact that skilled tradesmen and workers are not only needed, but lacking in areas, much to the detriment of our economy. These jobs can provide a very good income for a worker and their family.

One system that has worked in some states is offering scholarships to students to fill the gap of skilled professionals that communities need. The community usually offers a scholarship, and the student must show progress toward their degree to retain the scholarship, and will have signed on to work in the community for an agreed number of years. These programs have trained and recruited doctors, nurses, and lawyers for rural communities for years. Such a program should be expanded to include inner-city areas and any required profession that is lacking. That expansion can be funded with federal money run through local agencies that would need to comply with federal requirements to qualify for that funding. That way, society and the government are also getting a tangible benefit for the money being spent.

The local agency (local government, research institute, or local/federal program) would identify areas that need skilled professionals and then request financing from the Department of Education. The local agency would then interview and award an interest-free loan to a student to attend a public college with the agreement that upon graduation they spend four to six years working at a reduced salary in the needed area, essentially working off the loan. The agency would agree to pay the professional a reduced salary and pay the difference between the reduced salary and the market salary for the job to the federal government to repay the financing. That way the government gets all or most of the money back to help fund the next round of students. By doing this, the budget spent on college loans would eventually transform into a stand-alone fund without further federal budget dollars.

The amount recovered to the Federal Student Loan Program should be viewed as part of the decision process. Proposals from local governments and research institutions would indicate the severity of their need, the amount requested and the amount to be paid back, i.e., the return on investment. The loan program would then award them based on a set criterion showing the entire decision process and allowing public review of it. Local needs should also be considered. Areas without doctors or lawyers should get some consideration in the award process to help residents in that area. If upon graduation, a student decided they did not want to work in the local agency, they would accept the entire loan (which would be the tuition and any stipend that was included) at an interest rate set by the Federal Reserve and have to pay it off.

The only question is what happens if a student fails to graduate. Having the student assume responsibility for all costs up to withdrawing from college may strike some as the same problem we now face with student debt. However, the student must take personal responsibility for not achieving the college degree, but the local agency deemed the student as worthy, so they also need to shoulder some of the blame for encouraging the student—but not all. Some may think this is excessive, hitting the student with debt, but they went to college and did receive some benefit. Especially in the case of advanced degrees. For instance, if a student has earned a bachelor's degree but failed to complete the advanced degree, they still benefited by getting a bachelor's degree. As such, there should be a split in responsibility between the student and the local agency. The student should be responsible for 60% of the costs with the remainder to the local government. Why the local government? Because the federal government is not only burdened with a large national debt but would also require a massive bureaucracy to handle the function of matching the scholarship to the student and likely wouldn't do it well. If the local agency was matching need to student, then the local agency would be the responsible party for the student loan with federal financing. This would make the scholarships sought after, but also require that the local agency, on the hook for 40% of the costs if the student fails, be held responsible for the screening to determine if the student really had a chance to succeed. This is really a contract between the student and the local government/research institute with the Federal Student Loan Program funding the scholarship. The local agency may give the loan to local students, which is fine, with the local agency responsible for 40% if the student fails.

This means everyone in the process—the federal government, the local agency, and the student—all have a stake in the program's success. This would be geared toward providing services for communities (doctors, lawyers, teachers, and healthcare workers) as well as supporting research and development (mathematicians, scientists, and engineers) that the federal government wants to encourage.

There are existing programs that allow active-duty military and veterans to attend college or even transfer their benefits to their children. These programs are not being touched and are expected to continue to be funded. The existing programs and this new one won't totally get low-income children access to college, but it is a start, and it minimizes federal involvement except where it is being used to try and develop professionals for needed areas. Some may feel this is a callous approach, to stop scholarships and loans that encourage college attendance and replace them with a system that may border on indebted servitude. But really, with a massive federal debt, how long can we afford to give away federal money without a positive result and actual benefit for society. The existing Federal Student Loan Program has buried millions of Americans under debt when they failed college and cannot pay off the debt. To continue that same program while talking about debt forgiveness is not recognizing what made the problem in the first place and only treating the symptom. When the scholarship is a success, the local community gains a needed professional at the cost of room and board for a college student, and the student gains a degree and a career at a reduced salary, and the Federal Student Loan Program gains all or most of the money back to then fund more loans.

The remaining question is what to do about the existing student debt problem. The Center for American Progress looked at this and showed that there were options for a partial debt forgiveness and laid out a number of scenarios. They showed that forgiving up to the first $10,000 of every existing student loan (one scenario of many) would cost about $371 billion (less than 25% of the entire debt) and 16.3 million borrowers would then be debt-free. [34] Since the average debt on those defaulting is ~$10,000, it would greatly lower the potential for defaults and remove the burden from those with low incomes who were burdened with student debt. If people didn't agree to provide relief for those with high incomes, then an income threshold could be added, but my view is that adding an income threshold would complicate the discussions and remove some of the support for this proposal. Realize that with the state of the US budget, the $371 billion is going to the federal debt. While a major point is to hold or shrink the debt, this is one of the few issues where growth of the debt is the right answer.

With a few twists, this is pretty much what President Biden has proposed. There's just one problem: he has been blocked by the courts. And rightfully so. Student debt relief is not something that can be ordered through executive action. This is a congressional issue. Congress controls the purse strings and Biden's proposal transfers the debt from the students to the nation. The massive national debt is Congress's responsibility. It is up to Congress to not only resolve student debt, but to put in place the changes in the Department of Education to fund scholarships and stop the loan program that created the debt in the first place.

15. Early Education

It is easier to build strong children than to repair broken men.

—Frederick Douglass

Too much time and effort is being spent on remedial courses at the college level. Remedial courses are required for students to build up their skills in math, reading, or English before they can take regular college courses. A lack of preparation in our primary and secondary education is requiring these remedial classes or "developmental education" in colleges. A 2016 report from the Center for American Progress estimated that 40 to 60% of first-year college students need to take at least one remedial class.[35] This is putting a burden on the students and colleges when the issue should have been addressed in elementary, middle, and high schools. The same report estimated that the costs to the students and parents are somewhere on the order of $1.3 billion per year. They also reported that students taking remedial classes are less likely to graduate.

So, is the issue that colleges are allowing unqualified applicants, or that we aren't sufficiently preparing our students at the primary/secondary level? The previous chapter addressed the student loan system that was encouraging colleges to let in more students without a concern for costs. Hopefully that addresses the issue of college admissions. Student preparation is an issue, as shown by reports of the performance of American high school students compared with other countries. America repeatedly ranks low when compared to other developed countries. The Pew Research Center reported that tests that measure the reading, science, and math skills of fifteen-year-olds in developed and developing countries placed the United States at an unimpressive thirty-eighth out of seventy-one in the math testing and twenty-fourth in both science and reading testing.[36] This set of seventy-one did not include China, and China has consistently outperformed the United States in secondary school test scores.

In 1957, the Soviet Union launched Sputnik, which achieved an orbiting manmade satellite before the United States. While America eventually won the race to the moon, the shock of Sputnik caused funding for education to triple the following year. We haven't had our Sputnik event yet, but Russian hacking of American cyber systems and China's continued advancements in artificial intelligence and other technology should be clear indicators that America is falling behind, and we should do something now rather than wait for another Sputnik moment to make the point painfully obvious. In the previous chapter, we proposed changes to the federal funding of college education into a more focused, but locally managed effort to develop students through scholarships. The same effort is needed to help America's 97,568 public elementary and secondary schools in the United States to better prepare those students for college and life in general.

There are 32,461 private elementary and secondary schools in America, which are being excluded from this discussion because their students are 50% more likely to go to college, and their college graduation rates are over twice as high as public schools. That begs the question: why not use federal dollars to support private schools and send more students to private schools? There has been a strong push by conservatives for school vouchers and other means to use public money to support private schools, but that is more about defraying costs for existing private school attendees than encouraging more students to attend private school. There are three reasons that could explain why private schools tend to be more successful than public schools.

First, student-to-teacher ratios are lower at private schools (twelve students per teacher versus over sixteen students per teacher on average at public schools).

Second, parents of private school students are likely to be better educated. The National Center for Education Statistics showed that ~32%

of private school students have parents with a bachelor's degree or better compared with only 15% in public school.

Third, private students are less likely to live in poverty with only 8% in poor households versus 18 to 19% for public schools.

Private schools are focused in urban and suburban areas (89% of students) versus rural areas (11% of students). The overall trend is that private school students come from homes with better educations and resources than public students. The issue may be less about the quality of the school and more about the quality of children who attend each type of school. Growing evidence indicates that success of a student isn't affected by the choice between public or private when the data is adjusted for other factors. A study by the University of Queensland in 2015 found that four factors had the most impact on the success of a student: their birth weight, the amount of time a mother spends with her child, the education of both parents, and the number of books in the household.[37] And while this study only tracked 4,000 Australian children through elementary school, additional studies in the United States and United Kingdom confirmed these findings. Private school supporters are quick to point out the advantage of private schools in graduation rates, college success, and financial success, but these better outcomes don't recognize that public schools are starting out with the burden of more students from poor and non-college educated families who are not as prepared for school as their private-school counterparts. The research papers that have taken this into account all come to the same conclusion: it's not about the private versus public school issue, it's about the quality of time and education that children receive prior to school and at home.

Five- and six-year-old children are inheritors of poverty's curse and not its creators. Unless we act, these children will pass it on to the next generation, like a family birthmark.

—LYNDON B. JOHNSON

So, the solution for providing better students to college is about improving the students before they ever show up to elementary school. Then this really is about what happens at home, and how is that to be addressed? This leads to a long-standing social issue of how to break the cycle of poverty and show people a path out of poverty, no matter their race. But we can't have the federal government intervening in people's homes. The mechanism and means of how to help isn't clear, and no one really wants a governmental presence in their homes. The best option is to provide some equivalent to allow low-income families access to an environment where children can have access to books and education prior to public school. Democrats had submitted a bill to both the US Senate and House in 2019 that aimed to expand daycare for working families and ensure they didn't pay more than 7% of their household income on daycare, but it wasn't passed. An added issue is that most childcare available to working families is provided by very low-income workers. Wages in childcare average just $11 per hour. The need is not only to provide childcare but raise the quality of it where it provides the same benefits that successful students are receiving—access to reading and attention from adults to improve development. There is also an advantage for working families in providing this high-quality daycare with a decrease in cost, as some studies show that working families are paying 10% or more of their household income for daycare. Another issue is that at present fewer than half of students have pre-kindergarten available to them through existing public schools prior to entering kindergarten.

That first solution is to support providing quality daycare for children up to and including pre-kindergarten. There are existing programs to help poor families with childcare, but they are geared toward only those in poverty. The Head Start and Early Head Start program are federal grants to local programs that are run by local agencies, local school systems and charities in most cases. These programs serve just over a million children, but present estimates are there are 4 million children in poverty under age six. An article in the *Atlantic* in 2016 estimated that Head Start was only reaching 41% of three- and four-year-olds and less than 4% of those younger than three.[38] The article indicated that some local programs achieved amazing results, but that there was no consistency as other local programs are struggling. Not only that, but some local programs don't offer childcare five days a week for a full day, making it difficult for working parents.

This is not a case for more federal oversight; some childcare program administrators indicated that they spent far too much time documenting how they met federal requirements for Head Start. It is a case for increasing federal funding and establishing a way to track the benefit to the students and encourage the exchange of ideas from those programs that are succeeding. To expand these programs and allow middle-income families to attend at a reasonable cost to the family would also be a great idea. Middle-income families are struggling to find quality daycare to allow them to work, and expanding the program to allow them to enter at a fee could benefit both them and the program without burdening the government financially. This could even be expanded to allow people with larger incomes to participate at the full cost per child. That has the potential to increase the programs' size, reduce the cost per child, and foster diversity.

The cost? Presently we spend $10 billion on Head Start. To fully cover children in poverty would require increasing it to $30 billion a year. Think that's too much? A new US Navy aircraft carrier costs $13 billion, and there are two presently scheduled to be built. That brings to mind a classic bumper sticker :"It will be a great day when our schools get all the money they need, and the Air Force has to hold a bake sale to buy a bomber." We spend more on defense than the next ten largest countries combined. Before we have our Sputnik moment, it is time to make education a priority.

This is not advocating slashing the defense budget, but finding the additional money to expand Head Start and Early Head Start to provide more jobs in those programs, increase the pay for those jobs, and improve more children's lives. At present, we spend a trillion dollars on Social Security, we spend $400 billion on Medicare, $400 billion on Medicaid, $700 billion on defense, but only $10 billion on early childcare and education to break the cycle of poverty in our children.

For poor and middle-income working families, access to quality daycare is an issue, and it doesn't stop at age five. Some school systems offer afterschool care, but it is not available everywhere. An After School Alliance report in 2014 estimated that 10.2 million children were enrolled in afterschool programs, but another 19.4 million children would be enrolled if programs were available.[39] Not only would that be a relief on working parents, it also could benefit the children. The New York City Department of Education began an afterschool program called School of One in 2009 to help children with math. The result has been heralded as one of the one hundred Best Innovations of 2009 by *Time* magazine and has shown that students participating have made math gains at 1.5 times the national average. Increasing America's skills in math, especially for children of poverty, cannot be stressed enough. Every study and analysis that looks at people's paths out of poverty shows that education is the ladder out of poverty that is most cost-effective. The second-best way out of poverty? Dual earners, where both adults in a household have jobs, and the only way to do that is to have daycare or afterschool care.

The need is to encourage the expansion of afterschool programs especially in low- and middle-income areas. This should be done the same way that the Head Start program was expanded, supplying federal funds and combining them with a sliding scale of payment so people below the poverty line (~$25,000 per year for a family of four in 2020) pay nothing and middle-income parents pay something but not in excess of 7% of their take-home pay. Expanding afterschool programs to allow more students to participate and access afterschool education programs like Teach to One (School of One's name outside New York City) would provide a benefit to the students and working parents.

The last item to consider about childcare is that while the Child and Dependent Care Credit can be deducted on a person's tax form, the byzantine rules and calculations make it so the credit is extremely limited, whereas every expense that a businessperson incurs gets deducted from their income without reduction or question. We need to recognize that for working families, childcare is a necessary expense to allow them to work and should be deducted as such in place of credits and other items to lessen the impact.

Education breeds confidence. Confidence breeds hope. Hope breeds peace.

—CONFUCIUS

A surprising alternative solution is for facilities to combine a nursing home and a children's daycare. Or maybe it shouldn't be so surprising, considering that more than a hundred years ago, families tended to live in multigenerational homes where grandparents were a source of childcare to allow parents to work. America faces an issue with caring for the elderly and maybe, just maybe, we can solve two problems with one solution. A report from Generations United and the Eisner Foundation in 2018 about the existing 105 facilities that combined childcare with senior care in the United States showed that both children and seniors felt an improvement in their quality of life by these facilities.[40] There was one facility in New York where at-risk youth (students older than what is being targeted) interacted with a Geriatric Career Development Center where the students' progress after the program was tracked and 97% graduated from high school and 90% went on to complete college. This point really hasn't been fully correlated but does seem promising. The report is strong on opinion but lacking hard evidence. It indicates that there were savings from shared costs between the senior center and childcare facility. What is also noted is that high-quality staff were extremely important in making these facilities work.

The potential to combine childcare with senior care and assist those facilities in becoming more affordable for residents from low- and middle-income families seems like a winning solution, but the research and empirical data to determine how to do this is simply not available. Too often, good objectives have been stymied by poor implementation on the governmental level. A system similar to that proposed for energy infrastructure (in a later chapter) where proposals for establishment of senior/childcare facilities from local institutions would be reviewed and ranked on cost effectiveness and assistance to the communities might be the best solution. An initial policy of a federal funded budget from the Department of Education and Department of Health and Human Services to construct and start up such facilities is what is being proposed right now. This should be combined with a mechanism where the established facility pays back all or some of the start-up capital provided to finance future facilities. Proposals from local institutions would be evaluated and funded based on the cost effectiveness and their impact to low- and middle-income working families. Programs would be required to commit to pay all or, in the case of especially disadvantaged areas, some portion of the start-up capital back to the federal fund, which would be part of the proposal and evaluation. After five years, if the program has met its goals and can show both assistance to the seniors, the working families, and the students' success in elementary school, then 50% of the loan would be forgiven.

That may seem harsh, requiring a payback from the facilities, but the facilities should be collecting tuition from parents and some form of payment for providing housing/daycare for the seniors. A facility that can pay back is financially stable and has a better chance of performing and more importantly surviving for the long term than a facility that constantly needs subsidies from the government. These are the types of facilities we need to encourage and develop at first, the low-hanging fruit that will fund future facilities and not increase the federal budget burden. At no time and at no level can the existence of the federal debt be ignored. If the facility is in a rural or inner-city area where poverty is especially prevalent, then exceptions can be looked at, but the more financially stable that facilities located in poverty prone areas are, the more likely they will succeed in the long term.

This is not saying that public schools are not in need of help. There are massive differences in the funding of public schools between rich school districts and poor ones. In a 1973 decision (*San Antonio Independent School District v. Rodriguez*), a father of students in a poor San Antonio, Texas, school district sued claiming the unequal funding was a violation of the Fourteenth Amendment's equal protection clause. The Supreme Court ruled that "the right to be educated (as a child of school age or an uneducated adult), was neither explicitly nor implicitly textually found anywhere in the Constitution. It was, therefore, not anywhere protected by the Constitution." The school district where the father had been sending his child is still underfunded compared to others. The issue is that most school districts are funded by local property taxes, which means that the difference in home values causes a difference in funding per student. A subsequent 1993 lawsuit ruled that the unequal financing violated Article 7 (Education) of the Texas State Constitution, and Texas then implemented a redistribution plan where funds were taken from rich districts and given to poor districts in a complicated formula (Robin Hood laws). It hasn't equalized financing, and in wealthy districts, it is hated, and there have been repeated efforts to repeal it. Unfortunately, the Supreme Court decision bars the federal government from intervening and likely that is for the best. But it needs to be studied to determine whether there is a way to improve schools in poor areas. As we have seen in other chapters, poverty is a national problem and relates to the issues of crime, racism, and college graduation. Equalizing or at least helping poor school districts to achieve parity with wealthy ones is needed, especially if there is an effort to improve the students before they reach public school. There is no point in investing in preschool improvements if we are not going to address the actual schools.

The Early Education Policy:

- To increase the scope of Head Start and Early Head Start federal programs and extend them to all families with families not living in poverty paying a sliding scale of cost;
- Work to increase the availability of afterschool care for K through twelve students that can provide extended care for working families and to provide afterschool education;
- Allocate a small amount to provide for the construction and start-up of combined senior/childcare facilities. Combine this with studies to determine the effectiveness of senior/childcare facilities, and if they show results after five years, then begin to use more funds to encourage these types of facilities.
- Require that all school districts provide the funding per student both before and after they have implemented Robin Hood laws, and use this information to make recommendations to the states.
- And if we want to be really bold, propose a US Constitutional Amendment to make education a requirement.

16.
Social Media

The trouble with the world is not that people know too little; it's that they know so many things that just aren't so.
It's easier to fool people than to convince them that they have been fooled.

—Mark Twain

The proliferation of conspiracy theories has become a problem. A person now is more likely to believe an unfounded conspiracy theory if it corresponds with their worldview than a reality that doesn't.

Social media hasn't helped—posts, memes, and articles from dubious sources are liked and shared before anyone makes an effort to fact-check them, and it is rare that a correction or challenge to the original post is ever seen by as many people who saw the original post. The term *fake news* is unfortunately applied to both the fake and the real. Anonymity on the internet has enabled people to effortlessly lie without consequence. Public figures on all sides have even begun lying and misleading shamelessly. That willingness to lie to advance a cause and the public's short memory have allowed them to escape any stigma. On one hand, far-right conservatives claim they are being censored unfairly and want to take action against the tech giants. On the other, liberals want the tech giants to somehow clean up the mess of false statements and propaganda on their sites.

Posting on social media is not the same as free speech. The social media platforms are owned and operated by companies, and they have requirements and rules that users agree to before they can log on. The long legal document that defines those rules and requirements is something that most people accept and sign off on without reading and barely even try to scan, like so many other long legal documents stuck in front of us. But they were agreed to. If a user is banned or blocked from a social media platform, that is not a violation of free speech, it is a result of the user's violation of the agreement that the user accepted with the social media platform. So, any claims that the social media platform is denying free speech are irrelevant. Everyone is still capable of going out in public and voicing their opinions, but social media is not a public sphere. It is owned, managed, and funded by private companies, which means we should think of it more like a newspaper, radio station, or TV channel, where the users are writing the scripts, not as a public square down by the courthouse. The fact that political campaigns are increasingly turning to social media to spread messages and propaganda doesn't change the fact that this is not a space where freedom of speech is guaranteed.

Russia released massive propaganda in the 2016 US presidential election, but that is something that US spy agencies and other countries' spy agencies have also done in the past to affect other countries' elections. This isn't new, and it won't be stopped by crying to Russia, and it will certainly happen again. The only criminal act was the hacking of the Democratic National Committee emails, which showcased the private discussions about preventing Senator Bernie Sanders from becoming the Democratic candidate since they feared he would lose. The exposure of private communication to the public is something everyone needs to consider as a possibility, especially in this day and age. Your communications are private until someone hacks them and shares them with the world. Even before email, the potential for a letter to be intercepted and published always existed, and depending which side of the issue you were on, it was either the act of a patriot or a traitor. The true issue with Russian actions in the 2016 election was that they found cracks and fissures within American society that gave them material to work with. This, coupled with an amazing willingness among some Americans to swallow that propaganda whole without questioning it, created the perfect storm for the Russians. That does not mean that the 2016 election was stolen or irregular. It does mean that propaganda, both foreign and domestic, will always be used to flood media channels before an election.

It is not the social media companies posting propaganda. Something the Trump campaign effectively did in 2020 was establish a loose, informal organization of users who agreed with their views and then supplied them (directly and indirectly) with memes and images they could repost. Some images were cute, some disturbing, but all of them were geared at getting people to vote for Trump. Much of it was filled with misleading and sometimes outright false information. This was also done by other campaigns and individuals, but the Trump campaign is singled out due to their extensive use of third parties, which is sometimes referred to as a disinformation network. This network of individuals and several far-right online media outlets were happy to jump on the bandwagon and proclaim all things Trump. Social media companies have provided the platform, but it is the users, both real people and automated accounts controlled by software programs (bots), that are filling the platforms with content.

So, what should be done?

Anonymity allows bots to mimic real people, and it allows real people to act without fear of reprisal. Anonymity has also spawned the act of "doxing," where real personal information is shared online to encourage harassment. This has become commonplace and even escalated to "SWATing," where fake police calls are used to have a police SWAT team raid the victim's house. These have turned deadly in a few tragic cases.

The first thing to do is make all accounts verifiable. The tech companies need to determine who owns an account and that it is a verified person. Companies or government agencies can establish non-human accounts for themselves, but they also need to be verified. This may take a little work to determine the best solution, but if you have a Facebook account, Facebook knows it's you, knows you are a real person, and knows how to get in touch if you break the law using that account. A way to fine violators would also be helpful, such as a valid credit card, something so financial penalties are available if necessary. Further, if the form of payment is invalid or was allowed to lapse, then the social media company is liable for any fines. That may seem harsh to hold the platform accountable for actions by users, but it is to help the social media companies verify valid accounts. If you are concerned that this stifles free speech by having non-anonymous accounts, remember that social media platforms aren't free speech; someone is paying for the platforms, the software, and the computer servers hosting the platform. It provides a venue for free speech, but it cannot be seen as allowing anyone to say something on the platform without fear of knowing where it came from.

The First Amendment says:

Congress shall make no law respecting an establishment of religion or prohibiting the free exercise thereof; or abridging the freedom of speech, or of the press; or the right of the people peaceably to assemble, and to petition the Government for a redress of grievances.

Nowhere does it guarantee the right to anonymity or state there will not be consequences if your speech harms someone else.

Consider that any newspaper must verify ads before they are published, otherwise they could find themselves in violation of the law. One example was in a *New York Times* article in March 2006. Adam Liptak wrote,

If this newspaper were to publish a classified advertisement for an apartment rental that said, say, "African Americans and Arabians tend to clash with me so that won't work out," it would be liable for housing discrimination under the federal Fair Housing Act. Yet Craigslist.org, the enormous online forum, posted that very ad in July, and most legal experts say, as the law stands today, Craigslist bears no responsibility for it.[41]

That law is the infamous Section 230 of the Communications Decency Act of 1996, which gave internet providers blanket protection. Conservatives have been advocating for its removal. And maybe that needs to be considered, especially as newspapers and magazines don't enjoy the same protections.

Any change to Section 230 is something that social media companies are not going to like, since it interferes with the model where users get immediate feedback once they have posted something. Several dating websites have already started to delay changes to profiles or uploading pictures so they can be reviewed before they are published. Requiring all sites to review/check all posts prior to publication is going to require either some brilliant artificial intelligence (AI) programming or a lot of reviewers. Unfortunately, with repeated cases of bullying, doxing, and propaganda, it may be necessary. The dividing line between a funny political cartoon and a piece of propaganda is so blurry that AI may never be able to sort it out, and since I like a good political cartoon as much as anyone else, I'd hate to see the cartoonists of the world banned from posting their material.

But requiring social media companies to review posts prior to publication is going to cost them money and interfere with their business model where users don't pay a penny and the social media company gets to mine, analyze, and sell the data, and use it to publish targeted ads. That free access allows them to have millions of users who see those ads and generate all that data for mining, which in turn pays for the social media platform.

The solution is that while internet providers are not the creators of anything that violates laws governing published content, they are the publishers, and they are the only way for law enforcement to reach the source of the content. This goes back to removing anonymity and having a way of imposing financial consequences to an individual's actions. Now the ACLU has argued, "If Section 230 is stripped of its protections, it wouldn't take long for the vibrant culture of free speech to disappear from the web." This was in response to an effort in July 2013 by attorney generals from forty-seven states to remove the protections in Section 230.[42] But if it is possible to fine a person directly for publishing harmful information or false propaganda or by taking harmful action, does that truly limit free speech? Those same actions can have real-life consequences. What about doxing or bullying? Are those protected as free speech? Of course not! Without any consequences, the actions causing the problems are unlikely to stop, so empowering a law enforcement entity via social media provider to take action seems like a valid response.

Another idea would be to accept that an immediate response is out of the question. Like someone in a crowd who suddenly decides to scream an obscenity, you can't really stop them from doing that, but you can include a consequence. Have the social media companies agree that if someone finds a post that is false or offensive, not personally offensive, but overall offensive, they can flag and report it. By *personally offensive*, we mean that you may not like a joke, but the majority of the population may not find it offensive. What is offensive is going to be a gray area—something where some tolerance is required, as all individuals don't share the same views. At first, the threshold should be if something is demonstrably false or harmful. But we need to be careful. Moving too much into the offensive category risks intolerance that can and should be viewed as overreach. We won't try to work out all the details and degrees of offenses here, but maybe leave this to a committee of stand-up comedians, since the wrong decision could put a lot of material for jokes out of reach.

If multiple people (and that would be for the social media company to identify and most importantly publish) mark an item as offensive, then reviewers investigate, and if they agree it is offensive or an outright lie, they remove it and flag the account. How many strikes each user is allowed is part of the company's definition and published so everyone is aware of the rules. And they possibly assess a fine to compensate for the cost of the review. The reviewers are an overhead that burden the social media company, and since that burden was provoked by the action of posting the lie/outrage, then the user who posted it should bear some of the cost.

The solution will need to evolve over time, just as social media and the internet have evolved, but first steps need to be taken.

The Social Media Policy

- Revise the Section 230 provisions to require social media companies to verify accounts tied to a responsible legal entity (person, corporation, government body, etc.) and some means of payment for any fines that are incurred;
- Failure to have the form of payment would leave the social media company responsible for the payment;
- Require social media companies to define and publish their procedures to allow them to ensure that published content in violation of laws is identified and removed.

17.
Fake News

You are entitled to your opinion. But you are not entitled to your own facts.

—Daniel Patrick Moynihan

In an effort to fill the 24-hour news cycle and to be the first to air a story, mainstream media has aired bad or early information that is misleading. The effort to take the time to dig into a story and truly work through all the layers runs contrary to the get-it-out-there-first attitude. This is complicated by opinion shows masquerading as news. And news organizations are no longer located in brick-and-mortar buildings with broadcasting towers. The explosion of online and cable news channels has not only multiplied the number of news suppliers but allowed them to focus on niche audiences. If there isn't a news channel focused on people of a particular hair color, there likely will be in coming years.

The multitude of news sources highlights the problem of alternate facts. In an infamous *Meet the Press* interview on January 22, 2017, Kellyanne Conway made the statement that the Trump administration was giving "alternative facts." She later tried to indicate that she meant additional facts and alternate information. But there is no denying that the incident centered on the crowd size during the Trump inauguration in 2017. Sean Spicer, the press secretary for the president, misrepresented the crowd size, flying in the face of clear facts that contradicted him.

People can draw alternate conclusions, but the publication of false facts from the government or anyone claiming to be a news agency is an issue that needs to be faced.

Longtime newsman Dan Rather's response to the Kellyanne Conway statement was:

When you have a spokesperson for the president of the United States wrap up a lie in the Orwellian phrase "alternative facts".... When you have a press secretary in his first appearance before the White House reporters threaten, bully, lie, and then walk out of the briefing room without the cojones to answer a single question... Facts and the truth are not partisan. They are the bedrock of our democracy. And you are either with them, with us, with our Constitution, our history, and the future of our nation, or you are against it. Everyone must answer that question.[43]

The question is what to do about "alternative facts." In the case of the government, that needs to be left to the voters and the existing laws. In the case of news media, the American public needs to have some protection that a group identifying itself as a news organization is vetting the facts and confirming they are real and truthful before airing or publishing them. That is not to say that mistakes are not allowed, but that the news media needs to own up to their mistakes and publish/air corrections as required. But what to do about the organizations that refuse to acknowledge and even go out of their way to mislead? Wikipedia's entry on "Fake news websites in the United States" listed multiple fake sites showing how widespread this has become on social media.

In his book *Why We Did It*, Tim Miller describes how several right-wing news media sites repeat some stories from main sites, but also fed their followers with stories that were thrown together to amplify the views and beliefs of their supporters. That's not news, that is outright lying. That behavior from the right, left, or mainstream media should have consequences.

In the United States, news organizations enjoy protections, both the First Amendment of the Constitution and local "shield laws" that are designed to protect reporters' rights to refuse to testify about information and/or sources of information. Any organization that provides alternate facts is no longer a news organization and loses those protections. ACLU lawyers would probably wince at that last line, since they could see an authoritarian government strip valid news organizations of their protections to stop them reporting on the authoritarian government. Consider how many times Donald Trump called valid news stories "fake news." Even he and his administration weren't the first to be caught; there are far too many examples of it, the most infamous being the Pentagon Papers showing the cover-up and lying by the government over multiple administrations during the Vietnam War. As much as there is potential for abuse, something does need to be done. A mob stormed the Capitol because they had been misled in two ways—first that a vice president could overturn the election, and second that Trump had the election stolen from him by voter fraud. Those two lies led to deaths, mob violence, and what some are calling a crisis in American government and society.

Fake news and alternate facts would be a lot less effective if listeners were more skeptical and better able to recognize propaganda. While earlier, we advocated for a better educated public (see the previous chapters on education), it will be some time before that effort takes effect and even then, with people's willingness to believe whatever validates their own beliefs, education is not a surefire solution.

That brings us back to some efforts. The first is that all news organizations should register and define who owns them, who finances them, and where they are located. Whether they are broadcast, cable, or only on the internet, they need to register before they can say they are a news agency and claim the protections provided under the First Amendment. If they don't register and claim to be a news agency operating in the United States, then they fail to qualify for the shield laws and should pay the consequences of existing libel laws.

Second, there needs to be a way for the public to flag news articles as false and report them to either the Federal Communications Commission or another designated agency. The best would be a system that accepts multiple challenges to the same article to initiate a review by the agency. If the article is false, the news agency is notified and can respond or publish a retraction/correction. If the news agency hasn't registered or they fail to respond, then they can refer them for prosecution. Much like the libel laws, you can't publish something false about someone and not expect a consequence. While there is no legal entity being injured, the injury is to society by the continuous spread of alternate facts which are nothing more than lies trying to masquerade as news. There should be consequences if someone repeatedly registers a news agency as false and the articles are found to be true, and for agencies that repeatedly release false information even if they later issue corrections, especially if the corrections are issued at midnight or when no one is watching.

Now, if the news agency is reporting on government activity, we get into the issue of whether the government can use this power to reinforce a coverup and suppress a story. That is where the ACLU must be focused. The potential solution is in the federal agencies' validation of the story. There are several sunshine laws and freedom of information laws, and this is where the new policy needs to be crafted. It should allow the agency and news media personnel access to data to either verify or refute a story. The law should be set up where if the agency cannot provide the reporter access to refuting data without question, then the case can't be acted on. For example. if a reporter claims there are aliens in Area 51, then the validating agency and the reporter need to be able to travel through Area 51 and confirm there are no aliens. If the government isn't willing to do that, then it can't disprove the reporter's claim and cannot prosecute the news organization or the reporter. In the case of a government story, the government must be willing to show its cards in public, otherwise it doesn't have a case. This is getting into the weeds, but remember this is geared toward a news organization that claimed Pope Francis was endorsing Donald Trump for president, which the Pope didn't, and nonetheless this information was reported just before the 2016 election and seen by many as true.

The News Media Policy

- Require news media entities to register, so they can be recognized as news agencies for legal protection.
- Establish a way for citizens to report news agencies for releasing false information
- Once an established number of reports for an article is received, have that information verified, and if false, notify the news agency to correct it. If they fail to do so, prosecute them and have them be decertified as a news agency.

18.
The Secrets of Tort Reform

The minute you read something that you can't understand, you can almost be sure that it was drawn up by a lawyer.

—Will Rogers

Tort reform has been debated at length. Conservatives see it as a need to protect corporations and society from frivolous lawsuits, while liberals see it as an unnecessary protection to allow corporate abuses. The reality is somewhere in between. A tort is defined as a case where a wrongful act or infringement of a right (other than under a contract) leads to civil legal liability, i.e., if someone causes you harm outside a legal contract, you can sue them to compensate for the damages.

The most famous tort case may be the one where a little old lady in Albuquerque sued McDonald's because her coffee was too hot and got almost $3 million. That's the urban legend, but the facts are actually more complicated. The incident happened in 1992, and the woman, 79-year-old Stella Liebeck, was not driving, but sitting in the parking lot in the passenger seat of a car driven by her nephew. In the act of removing the lid from the coffee, braced between her legs to add cream and sugar, she spilled it in her lap and suffered third-degree burns (burns where the full thickness of the skin is destroyed) to her pelvic region, requiring extensive medical treatment including skin grafts. She offered to settle with McDonald's for $20,000 to cover her medical expenses and lost income. They refused and only offered $800, which Mrs. Liebeck's medical bills had already far exceeded. During the trial, McDonald's admitted that their operating manuals required the coffee to be served at "195 to 205 °F and held at 180 to 190 °F for optimal taste." Between 180 and 190 °F is a temperature at which doctors testified that skin damage occurs within three to seven seconds of exposure. It also came out that between 1984 and 1992, McDonald's had 700 cases in which the coffee had burned other people and, in some cases, very severely. In some of these cases, McDonald's had compensated the injured party. The jury awarded Mrs. Liebeck $160,000 in damages and then assessed $2.7 million in punitive damages. The amount may seem excessive, but McDonald's witnesses admitted that the coffee was too hot to drink, and one witness tried to downplay the incidents saying that 700 cases was insignificant compared to the billions of cups of coffee McDonald's served over the same time. The judge in the case reduced the punitive amount from $2.7 million to $640,000. Mrs. Liebeck and McDonald's settled outside of court for an undisclosed amount and required Mrs. Liebeck to sign a non-disclosure agreement so she could never discuss the settlement or the case in public again.

This rendered her unable to dispute what followed. McDonald's used public relations to mount a media campaign that resulted in ABC News at one point calling the case "the poster child of excessive lawsuits." Not to pick on ABC News since many news outlets, politicians, and others took up the same chant during the 1990s. This resulted in several state legislatures passing caps on punitive damages in multiple states. A documentary *Hot Coffee* was released in 2011 and focused on the debate about limiting punitive judgments in civil lawsuits.

But, of course, McDonald's must have lowered the temperature of their coffee. Right? No, they didn't. McDonald's current policy is to serve coffee between 176 and 194 °F. And they aren't alone in that. Starbucks sells coffee at 175 to 185 °F, and the Specialty Coffee Association of America recommends a temperature of 160 to 185 °F (71–85 °C). In fact, most coffee suppliers provide coffee at temperatures greater than 170 °F. Several lawsuits have been brought since the McDonald's case and most have been rejected when the supplier of the coffee shows they are adhering to the industry standard. Remember, the McDonald's executive indicated that billions of cups of coffee were served, and only 700 cases had been an issue, which means that billions handled the coffee safely without getting burned, so why would the 700 have a reason to sue? The probability of injury being 700 among billions compares to driving, where death via motor vehicle accident consistently ranks as one likely cause of death in America. The National Safety Council placed the chance of dying in a motor vehicle crash at one in 106 over a person's lifetime, far riskier than any other activity. The only things more likely to kill someone were suicide, overdose, cancer, and heart disease. Almost 40,000 people died in motor vehicle accidents in 2018, and that is even with the fact that driving has become thirty times safer between 1913 and 2018.

If we view driving a car as an acceptable risk, then how can we view the risk of being burned by coffee as unacceptable? Especially if the restaurant serves the coffee at the same temperature as everyone else, and the odds of getting hurt are a lot smaller than the risk of driving a car. Remember, Mrs. Liebeck's injuries were so bad because she was sitting in a car and spilled the coffee into her lap and could not get out of the car before serious damage had been done. Was the coffee the problem, or was it how she was holding it between her legs as she took the lid off? Were her actions unsafe? This is not to say that there are not lots of examples where a manufacturer supplies a clearly unsafe product. But what is the level of consumer knowledge and personal responsibility that can be assumed by suppliers and manufacturers reasonably?

Bought a ladder recently? Take a look at all the labels on it about not standing on the top rung, not climbing up the back of the ladder, and how to use it safely. That may seem like a lot of unnecessary labels. Many ladder suppliers have been sued by consumers who use the product in what most people consider an unsafe manner. The result of those lawsuits? All those labels up the side of the ladder. Coffee cups from McDonald's, Starbucks, and everywhere else are now printed with a very large clearly worded statement on the side: "Caution: Contents Hot."

Consumers need a way to sue when manufacturers supply defective products and cause them injury, but manufacturers need to be protected from lawsuits when consumers use their products in an unsafe or unreasonable manner. So, should the few cases where people mishandled coffee be allowed to sue or should there be some standard of reasonableness? The answer is found in the existing US court system. The court systems are presently working very well and there seems to be little need to address the issue of frivolous lawsuits. If the parties don't find redress at the local court level, appeals courts offer a higher authority, and they can appeal to the supreme courts of both the states and finally the country.

In fact, the so-called litigation explosion feared by many and the justification for conservative tort reform hasn't really occurred. Tort cases (injury and malpractice) have declined. A *Wall Street Journal* article in 2017 indicated that "fewer than two in 1,000 people—the alleged victims of inattentive motorists, medical malpractice, faulty products, and other civil wrongs—filed tort lawsuits in 2015, an analysis of the latest available data collected by the National Center for State Courts shows. That is down sharply from 1993, when about ten in 1,000 Americans filed such suits."[44] This is confirmed as tort cases declined from 16% of civil filings in state courts in 1993 to about 4% in 2015, a difference of more than 1.7 million cases nationwide, according to an analysis of annual reports from the National Center for State Courts.

That does not mean that everything is fine. There are issues that need to be addressed.

The ability to sue a foreign manufacturer is complicated, as reported in the *Emory Law Review*:

> The current legal system makes it virtually impossible for US plaintiffs to sue [foreign] manufacturers, legal redress is limited or nonexistent for the majority of these consumers. Three primary procedural hurdles—personal jurisdiction, service of process, and enforcement of the judgment—prevent suits against [foreign] manufacturers. While

> injured consumers may bring suit in the United States against importers, distributors, and sellers of the defective [product], many of these potential defendants are out of business, bankrupt, or possess insufficient assets to satisfy a judgment. As a result, many consumers are left with no one to sue and thus no compensation for the harm they incurred.[45]

In the continuing issues with foreign manufacturers, this ability to avoid the US legal system provides them with an unfair advantage, giving them the ability to undercut prices of American competitors on top of the advantages they already have with lower employee wages. The solution proposed in the Emory Law Review was a consumer settlement fund modelled from the fund set up in the wake of the Deepwater Horizon accident. The management of the fund was based on examples of the Deepwater Horizon and the EPA Superfund. The problem was it was financed by penalties to foreign manufacturers after an incident. The problem is that this doesn't recognize that manufacturers can use shell companies or otherwise revise their names to avoid penalties, and some countries will be averse to enforcing US penalties and likely to assist their manufacturer in avoiding them.

The solution is to establish a consumer protection fund as part of the Bureau of Consumer Protection and to use a small tariff to start it. If a claim is made against a foreign manufacturer, the manufacturer and their country's representatives are notified and asked to attend the court case. If the manufacturer fails to attend or the verdict is against the manufacturer and the manufacturer does not acknowledge or settle the debt, then the foreign country is notified to settle the debt or appeal the lawsuit. If they fail to do either, then the fund will settle the debt, and the tariff on that specific country increases. If the fund for a specific country reaches an amount equal to 1% of the value of the trade being imported from the country, then any additional funds will be transferred to the US Treasury and used to paydown the national debt. If the foreign country agrees to hold its manufacturers liable and part of the US lawsuits, then after two years of compliance, the tariff will be reduced to its original amount.

There is another issue. The United States uses what is called the American Rule, where each side of a lawsuit has to pay their own lawyers and court costs, no matter if they win or lose. Whereas nearly every other Western democracy uses the English Rule, where the loser pays the winner's court costs. Some argue that loser-pays rules would deter marginal lawsuits and tactical litigation and create proper incentives for litigation. The supporters of the American Rule argue that it is necessary to allow small or individual plaintiffs to sue large corporations, the idea being that David isn't going to go up against Goliath if he knows he has to pay Goliath's lawyers as well as his own if he loses.

But that misses the point. In the present world, David isn't paying his lawyers, private equity firms are. As reported in the *San Antonio Express-News* in 2018,

> By some estimates, more than one-third of US law firms used litigation financing in 2017, up from 7 percent four years earlier, while private equity and hedge fund investments in lawsuits have surged to about $30 billion from $1 billion in 2011.[46]

We need to recognize that litigation financing was happening long before this. For a long time, many law firms would take on cases with a signed agreement from the plaintiff that they got some portion of the compensation (50% is a common percentage) in place of charging fees to individuals who couldn't afford them. And now, in addition to law firms doing it, this type of risky investment is attracting private equity for one simple reason: it pays big dividends. If anything should have Goliath scared to death, it's the idea of a lot of Davids who can sue without any personal risk backed up by a lot of guys in dark suits with suitcases of cash at the ready.

Changing to an English Rule system, where the loser has to pay the winner's court costs, really doesn't change the model of litigation financing, except that the losing side becomes more expensive and therefore the investor is only going to back cases that have merit. This may seem unfair to the small or individual plaintiffs, but where they were getting litigation financing already, it's no change at all, it just means they had to have a strong case to begin with. The risk to the litigation financier has gone up if he must pay the corporate legal costs, but so has the profit if the corporation has to pay his legal costs as well. For the Goliaths of the world, it means the "frivolous" lawsuits should stop, but it doesn't remove their risk or exposure when they are negligent. Since this is unique to each state and the federal legal system, it would need to be changed in each state and at the federal level.

The Tort Reform Policy

- Establish a compensation fund for lawsuits against foreign manufacturers and financed via a tariff that is increased on any country not complying with US lawsuits against the country's manufacturers.
- To work to change the US legal system from where each side pays their own legal costs to a loser pays winner legal costs for all civil tort cases.

Section IV: Energy Policies

Energy (noun): power derived from the utilization of physical or chemical resources, especially to provide light and heat or to work machines.

—Oxford Languages

Energy and persistence conquer all things.

—Benjamin Franklin

19.
The Realities of Coal

My administration will put our coal miners and our steel workers back to work.

—Donald Trump

In 2019, the United States consumed 586,539,000 short tons of coal and almost all of it (538,601,000 short tons, 91.8%) was used by electric power plants. That compares with 2004 to 2005 when coal consumption was over a billion short tons per year. Coal production is almost half of what it was fifteen years ago, and coal is down to one primary consumer. So, this means that the health of the coal industry is reflected in the health of the electric power plants that rely on coal.

The last new coal-fired power plant was built by the University of Alaska in 2019. The seventeen-megawatt coal-fired power plant is one of the smallest and is solely to supply power and heat to the campus of the University of Alaska located in Fairbanks, one hundred miles from a local coal mine. It was determined to be the best option to replace an aging system, since wind isn't a consistent power source and because they are almost in the Arctic circle, solar power isn't effective either. Natural gas and oil were considered but aren't available locally and would have required extensive pipelines or trucking to supply the power plant. As a result, coal was chosen as the energy source. Prior to this, the Dry Fork Station (422 megawatts) near Gillete, Wyoming, was the last new industrial coal power plant opened in 2011.

Most of the US coal power plants were built prior to 1990 and tend to emit twice as much CO_2 compared to natural gas-fired power plants. Coal-fired power plants emit on average 2.5 pounds of CO_2 per kWh, while natural gas plants emit 1.3 pounds of CO_2 per kWh. That difference is based on the chemical differences between methane (which is over 85% of natural gas) and coal. The energy generated by burning hydrocarbons comes from breaking the hydrogen-to-carbon bonds that exist in methane and coal. There are four hydrogen bonds in methane per carbon atom. In coal, there are a little over two hydrogen bonds per carbon atom. That is why you always get twice as much energy per carbon atom in natural gas as you do from coal.

That basic difference between natural gas and coal, combined with the age, inefficiency, and pollution issues of the aging coal-fired plants have driven a decrease in the number of coal-fired power plants, which have fallen from 1,024 facilities in 1997 to 241 in 2019. Electricity production from coal is half of what it was during the height of production in 2005.

In 2016, President Trump proclaimed he would bring back the coal industry, but the reality is that the coal industry's sole customer is coal-fired electric power plants. The continued decline in the number of power plants has been driven by economic considerations that a US president is helpless to change. The hard facts are that coal-fired plants are old, inefficient, and they pollute more than natural gas–fired plants. This is something that no one can change without a massive interference.

Another problem is that coal-fired plants have to be set up as base-load units. *Base load* refers to power plants that are set up to run at or near maximum capacity year-round because they are more efficient and economic when operated that way. Peak load or intermediate load refers to power plants that can be ramped up or slowed down to match electricity demand.

Lacking the ability to operate as peak-load units, an aging infrastructure, being more inefficient than other options and some of the largest polluters in the energy portfolio, has spelled the demise of the coal-fired power plants. That hasn't stopped politicians from proposing all kinds of incentives and handouts to try and turn this trend around. But this is a classic example of the government trying to pick winners and losers instead of allowing economic drivers to find the best solution.

The reality is that coal mining is a declining industry, and there is no reason to spend federal or state funds to revitalize it. Especially as modernization and automation of coal mines has already massively reduced the workforce needed for the coal industry.

In 1985, 178,000 employees worked in coal mines to produce 883 million short tons of coal for the year. By 2005 1,131 million short tons was produced with only 74,000 employees—a 28% increase in production with an almost 60% reduction in workforce.[47] Thanks to mechanization and automation, the amount of coal produced per man had tripled. In 2019, the workforce was approximately 52,000 employees and production was 706 million short tons of coal (this is production; US consumption was 586 million short tons; the US exports the rest). Coal production for 2021 was 577.4 million short tons, a 7.4% increase over 2020 when the pandemic limited production, but a sign of the continued decline of coal production.

Coal reserves within the country are fairly large. The Department of Energy estimates recoverable (what can be dug up) reserves at 250 billion tons of coal, which is why some people don't want to give up on it. But most of those reserves are in Montana, Illinois, and Wyoming (60% in these three states) and only a little bit (less than 10%) is at existing coal mines. Which is why there has been a push for clean coal, a way to use coal by reducing the pollution generated and capturing the carbon dioxide generated, but that hasn't worked yet. There have been substantial investments in clean coal, and so far, there have been very little results. Faced with climate change and having not achieved any technological breakthroughs to achieve clean coal, the future for coal is not looking good.

That is not to say that we should turn our backs on coal. Continued investment in research and development should be encouraged due to the large reserves that the United States possesses. But barring some technological breakthrough, the federal government doesn't have any business trying to change the basic economics of any industry. The government has done a horrible job in the past of trying to pick winners and losers in industries. Federal investment should be very small, due to the efforts to move to less polluting forms of energy and develop better, greener transportation systems.

Coal is fading for reasons driven by factors outside its control. To artificially try and change those factors would be to spend resources that could be best applied to other areas. The rural areas where coal mines were the major employers in the past wish for a return to the good times. While that is understandable, these areas would be better served by facing reality and spending money on education to retool for new industries, as opposed to longing for the return of days past.

20.
America's Energy Question

Earth provides enough to satisfy every man's needs, but not every man's greed.

—Mahatma Gandhi

Due to climate change, America needs to shift from hydrocarbon-based energy sources to more renewable energy, and that has been happening to a limited extent. This isn't a new idea. The educational series *Schoolhouse Rock!* debuted an episode called "The Energy Blues" in 1979, and it recognized that energy conservation was needed. Now, to put this in context, the United States was a major oil importer in the 1970s and had just come out of an oil embargo by the OPEC states that had driven oil prices to an unheard-of price of $11.65/barrel in January 1974. Fifty years in the past seems like a completely different world.

But there are present-day issues that need to be confronted and facts that must be considered. These facts are provided by the US Energy Information Administration of the Department of Energy, who tracks the energy industry and generates US energy statistics on their website (eia.gov). The website uses quads as energy units for yearly energy consumption or production. A quad is a quadrillion British thermal units. As discussed in the chapter on climate change, a British Thermal Unit (BTU) is the heat needed to raise one pound of water by 1 °F and a quadrillion is a one followed by fifteen zeros or 1,000,000,000,000,000.

Quads serve as a generic energy term to translate between barrels of oil, cubic feet of natural gas, and kilowatt-hours of electricity. Total US energy consumption was 100.2 quads in 2019, and we are using 2019 as we want to look at pre-COVID energy use.[48] The United States consumes 15.7% of the world's energy with only 4.3% of the world's population. If that seems like a massive amount of energy, it is. And that is the first thing that needs to be understood. The US energy industry is like a giant oil tanker at sea, huge and massive. It doesn't change course very quickly. That doesn't mean that it isn't willing to change. The energy industry has been evolving ever since it first came into existence. As the facts will show, the energy industry is a massive number of facilities located throughout the United States, producing heat and carbon dioxide that somehow needs to be converted to renewables if we truly want to be carbon neutral. But it is also the habits and patterns of the American consumers that has an influence on how the energy industry evolves. And those habits will need to change as well.

ENERGY SOURCES

- Liquid Petroleum: 36.7 quads (37%)
- Natural Gas: 32.1 quads (32%)
- Coal: 11.3 quads (11%)
- Nuclear Power: 8.5 quads (8%)
- Renewable energy: 11.5 quads (11%)

ENERGY CONSUMPTION (AND MAJOR SOURCES)

- Transportation: 28.2 quads (37%)
  - Liquid Petroleum (91% of consumption)
- Industrial: 26.3 quads (35%)
  - Natural Gas (40% of consumption)
  - Liquid Petroleum (34% of consumption)
- Residential: 11.9 quads (16%)
  - Natural Gas (44% of consumption)
  - Electricity (41% of consumption)
- Commercial: 9.4 quads (12%)
  - Natural Gas (39% of consumption)
  - Electricity (49% of consumption)

If you noticed that the number of quads in sources and consumption doesn't add up, this is because 37.1 quads of fossil fuel is used to generate the 12.8 quads of electricity. And yes, 65% of the fossil fuel sources are wasted generating the electricity. The major loss is the conversion of hydrocarbons (oil, natural gas, and coal) to heat and then to make electricity. Coal and oil power plants tend to be 30% to 44% efficient, meaning that 56% to 70% of the energy of the fuel is lost as heat up the smokestacks. Some natural gas plants can achieve greater then 60% efficiency, but this is usually done by recovering some of the heat in the exhaust and using it to heat homes or industrial processes nearby. Recovering heat and reusing it is called waste heat recovery and only works when there are nearby users who can use the heat.

So, people were talking about the Green New Deal proposed in 2019 by Democrats. The key elements were to reduce net carbon emissions to zero by 2050, in less than thirty years. There are other parts that will be discussed later, but let's look at carbon emissions first.

As indicated earlier, the United States gets 80% of its energy from hydrocarbon sources, which are the major carbon emitters. The idea is to replace that with renewables or develop carbon-capture to remove CO_2 from the atmosphere and store it/convert it. Carbon capture breaks into two technologies, capturing carbon dioxide at the sources (fossil power plants/industrial emitters) and direct air capture. Existing CO_2 capture uses amine processes to collect CO_2 from power plants and inject it underground into nearby oil fields to help oil production. When you look at these power plants, they are more expensive and require more energy per kWh than conventional plants.[49] For a coal power plant, a conventional facility without carbon capture is $3,676/kW of capacity and uses 8,638 BTU/kWh of energy. Turn it into a facility that captures 90% of the CO_2 output, and now it costs $5,876/kW to build and uses 12,507 BTU/kWh of coal to make electricity. That increased the capital cost by 60% per kW and the energy usage per kWh by 45%. The less efficient use of coal and higher capital cost make these facilities economically challenged. The existing ones that have been installed were only economical when considering the increased oil recovery that they provide by injecting CO_2 into depleting oil fields. The cost of adding carbon capture to natural gas power plants is almost as expensive, a 12% increase in fuel usage per kWh and a 128% increase in cost per kW to build.

Direct air capture is pulling CO_2 out of the atmosphere, but not in conjunction with power plants or industrial emitters. A September 2022 report from the International Energy Agency indicated that only 0.01 million tons per year of CO_2 was being captured by the eighteen existing plants but talked about scaling up to 60 million tons per year of CO_2 by 2030 to meet the net zero scenario plan.[50] And the problem is not only building out a massive increase of direct air capture (DAC), but there are also the energy requirements. Estimates are that DAC requires six to ten gigajoules per ton of CO_2. That's 1,666 kWh to 2,778 kWh per ton. Want to collect a million tons of CO_2 per year? Worst case that is 2.7 TWh of electricity. Say you are getting electricity from those coal power plants with 90% CO_2 capture, they would capture 90% of the CO_2, but still release 357,868 tons of CO_2 while capturing the 1 million tons with your direct air capture units. And we are not considering all the steel, copper, and other materials to build the DAC units and mine all the coal to get to that point. And if that electricity had come from, say, normal coal power plants it would have generated 3.6 million tons of CO_2, far exceeding the carbon captured by DAC. If the electricity had come from the most efficient natural gas combined cycle power plants without CO_2 capture, it would have generated 1.04 million tons of CO_2. Direct air capture combined with existing fossil fuel power plants is a losing proposition.

We are going to focus on replacement now, because carbon capture technologies are still in the developmental stage, and the reality is if you can replace kWhs of electricity with renewables, it is better than trying to install direct air capture to offset those same kWhs, especially for coal. Capturing CO_2 at fossil power plants and storing it underground may be desirable as well, but even then, the application will require large amounts of energy and a new infrastructure, which adds to the overall energy consumption. Several technologies are in development but are simply not to the point where policy can be based on them.

The easiest thing to look at is electric power generation, which is electricity generated and used in terms of terawatt-hours (TWh), one trillion watts of electricity for one hour, and in terms of megawatts (MW) of installed capacity, the ability to generate a million watts of electricity at one moment. Run 1,000 megawatts of capacity for 1,000 hours, and you get a terawatt-hour. The United States has 1,084,370 MW of power-generation capacity, but some of it doesn't run all the time and some of it (wind, hydro, and solar) are limited by environmental conditions and are less efficient. Generally, the United States runs nuclear (98,119 MW) and coal (about 280,000 MW) at full throttle all the time to generate base load, as coal and nuclear can't adjust well to changing demand. Wind and solar produce power when they can. They depend on changing weather conditions. Hydro and natural gas make up the difference in peak load, meaning they generate more or less electricity based on consumer demands. The changing load from consumers is something we need to address later, which has an impact on the desire to become a greener electric country.

The United States in 2019 generated 4,113.5 terawatt-hours (TWh) of electricity, which was split into 38.5% natural gas, 23.5% coal, 19.7% nuclear, 7.2% wind, 7% hydroelectric, and 1.7% solar.[51] In 2019, 6,181 megawatts (MW) of wind power generation capacity was added and 4,278 MW of solar power generation was added out of a total increase of 17,970 MW for the year. Natural gas was the big addition at 7,339 MW, but still the bulk of the new additional power generation was renewable. A megawatt of power generation capacity doesn't directly convert into actual terawatt-hours of electricity generated. Every energy source has an efficiency, and they aren't constant.

Let's look at the four basic types of renewable energy, and we will also cover nuclear, because no single type of energy generation will solve the problem.

HYDROELECTRIC

Hydroelectric generation is the oldest renewable power source used by the United States, and to generate it, all one has to do is find a flowing body of water (river, stream, brook) and build a dam across it, preferably where the dam can be very tall. Electric power is generated when the water flows through a turbine that is connected to a generator. The amount generated depends on the amount of water flowing through the turbine and the height of the water behind the dam. The taller the water level on the upstream side of the dam, the more electricity can be generated per gallon of flowing water. The problem is that this depends on the terrain and how the river is used.

The Mississippi River is the largest flowing river in America, but there are no hydroelectric dams on the Mississippi except in Minnesota, which means that for ~2,000 miles of its length, it is without hydroelectric power generation. The reason is that the river is a major waterway carrying millions of tons of barge and ship traffic, and it mainly flows across very open, flat land. Using it for hydroelectric power would be possible by installing small hydropower units that use propellers supported from bridges or other structures in the waterway, but a major dam like the Three Gorges in China or Hoover Dam in Nevada would be impossible since the surrounding terrain is too flat. The other issue is that dams interfere with the flow of a river, disrupting downstream users. Silt and organic material collect on the upstream side of the dam, creating other problems. The United States removed ninety dams in 2019, some of which were hydroelectric generators, to restore rivers. Environmentalists are trying to remove more.

The United States generated 287.8 TWh in 2019, or 7% of all US electricity, from approximately 79,800 MW of installed capacity at over 2,200 locations. That works out to about 41% efficiency of converting the MW of installed capacity into TWh of generated electricity. The reason the efficiency is not higher is that hydroelectric power depends on water flowing downhill, and that isn't the same year-round due to changes in snowmelt and rainfall.

The US Department of Energy in a July 2016 report estimated that 13,000 MW of hydroelectric capacity could be added to the network.[52] That would generate an additional 46.9 TWh per year, but environmentalists have fought new hydroelectric power plants and caused some older facilities to be removed and not replaced due to their impact on rivers. As such, we will assume the United States can get half of the predicted power for an additional twenty-three TWh from hydroelectric.

The report also indicated that 52,000 MW of pumped storage hydropower could be added to the existing ~23,000 MW of pumped storage. Pumped storage is like a battery where water is pumped uphill into a lake or some form of water storage when excess electricity is available, and when the power grid needs additional energy, the water flows downhill through a hydroelectric power plant generating electricity. These types of systems are not as efficient as batteries (15–25% lost energy for pumped storage versus 10—20% for electric batteries), but they last for years, don't lose storage capacity like a battery does, and don't require rare metals (lithium, etc.). With additional solar and wind power in the future, additional energy storage will be a requirement, so any future Green Deal needs to look at adding pumped storage for largescale energy storage. The problem is these units need large height differences between the lakes, so they will likely be located in the mountainous parts of the country and unable to support most areas.

There are some new river turbine designs that could be installed on flowing rivers like the Mississippi to generate small amounts of hydroelectric power and not interfere with navigation, wildlife, or the natural flow of the river. More research and trials are needed to determine how efficient and cost-effective they would be, so at this time, they are not included in this analysis.

WIND

Gone is the old frontier wind turbine with lots of blades in a circle and a weather vanning tail and in is the mammoth, three-bladed electric power generators of today. The large wind turbines used in utility-scale wind farms have blade lengths between sixty and 260 feet long and usually produce 1.3 MW or greater electricity per wind turbine. Some fifteen MW giants for offshore use are in testing as of the writing of this book. Smaller units that can be installed on houses and light poles are also available, but they all have the same issue: the wind doesn't blow consistently, and so they have to be married with some form of power storage.

In 2019, the United States generated 294.9 TWh from wind or 7.2% of US electricity, from approximately 99,663 MW of installed capacity. That works out to about 34% efficiency of converting the MW of installed capacity into TWh of generated electricity. No surprise—the wind doesn't blow all the time or at the same speed year-round. This is wind's major disadvantage—intermittent power. There has been some noise about how many birds wind farms kill, but that needs to be taken with a grain of salt, especially when compared to the number of birds killed by tall buildings and existing fossil fuel power plants.

Wind has proven to be very successful, as stated by *Bloomberg New Energy Finance* in April 2020:

> Solar PV and onshore wind are now the cheapest sources of new-build generation for at least two-thirds of the global population. Those two-thirds live in locations that comprise 71% of gross domestic product and 85% of energy generation."[53]

But the intermittent issue needs to be addressed, and that may be resolved if electricity can be moved across the country more efficiently, but most likely it will require largescale energy storage.

If the United States added wind power at the same rate as 2019 for thirty years, that would add 548.7 TWh of electricity generation per year by 2050. One of the concerns is that most prime locations for wind are being utilized or slated for development right now. As more and more units are added onshore, they may become less efficient. The United States has only one operating offshore wind development, but that is an area for growth that needs to be pursued as well. But it needs to be accepted by all Americans that the "not in my backyard" (NIMBY) attitude has to stop. Blocking an offshore wind farm to preserve the view from their beach house directly conflicts with reducing carbon emissions. We can either have pristine views or green electricity, but not both.

Another issue to consider with any of these generators is what to do about end-of-life. No electricity generator lasts forever, and photos of old fiberglass wind turbine blades being buried in a landfill have become extremely popular among the non-renewable community, but that fails to recognize the required replacement of parts in any form of electric generation, including all the fossil fuel ones.

Solar

There are two types of utility-scale solar (5MW or greater): the first is large arrays of photo-voltaic cells that use chemical/physical reaction of sunlight striking light-sensitive material in the cell, and producing very low voltage direct current. By combining multiple cells and then using electrical equipment to generate higher voltage and alternating current, AC power can be supplied just like what comes from the existing fossil fuel power plants. The United States generated 68.7 TWh from large utility photovoltaic installations of approximately 35,710 MW of installed capacity. That works out to about to 22% efficiency of converting the MW of installed capacity into TWh of generated electricity. Solar has the lowest efficiency of the renewables because it only gets sunlight for eight hours during the hottest part of the day with unobstructed daylight, as opposed to a fossil fuel plant that can run 24 hours a day. There is no issue with the efficiency of the solar cell; it is the availability of sunlight. Snow, dust, or leaves blowing onto the solar cells also reduces the amount of electricity that they will produce. In addition to large utility-sized facilities, there are several small and residential installations that amount to about 70,000 MW of installed capacity. These small installations aren't monitored or tracked, but the EIA estimates that they generated about 140 TWh, resulting in an efficiency of 23%. Something to remember is that photovoltaic panels have about a twenty-to-twenty-five-year life, so not only does conversion to solar power need to be looked at, but also the maintenance of these panels and how will they be replaced and recycled.

The other type of utility-scale solar power is concentrated solar power (CSP) facilities, where arrays of mirrors focus the solar energy on a central point to generate heat to make electricity through steam turbines or a thermochemical reaction. The CSP facilities can store heat for use during the night to generate electricity, thereby letting them work as baseload units for more constant electricity production. At the moment, less than 5% of utility solar installation are concentrated solar power because they tend to be more expensive to build. The United States generated 3.2 TWh from large CSP installations of approximately 1,756 MW of installed capacity. CSP is one of the more expensive renewable options, but that is because very few facilities have been built in the United States. CSP facilities are only going to be feasible in the desert regions of the southwest, where solar radiation levels are very high and open, flat land is available. A CSP installation needs to be about one hundred MW or larger to be effective and that requires 500 to 1,000 acres of land for the large array of mirrors to collect and focus the solar energy on the tower where heating takes place.

One of the advantages of solar is that it is almost in sync with the changing electricity demand. It produces the most power during the hottest part of the days during the hottest periods of the year. Solar production in December and January is less than half of what is produced in June and July.

In 2019, the United States added 4,278 MW of large utility-scale power generation to the national grid. If that trend continued, then the country would have an additional 265.6 TWh in 2050. Small-scale and residential solar has been growing by about thirteen to sixteen TWh per year, so we can hope to continue to achieve an addition 390 TWh of "small" solar.

GEOTHERMAL

Geothermal power works by sourcing hot water heated deep in the earth's crust to power steam turbines to generate electricity. This is done by either pumping water down into hot rock formations or relying on natural flows to generate the hot water/steam. Geothermal presently generates about 15.5 TWh of power generation from utility-scale installation of 2,535 MW. Geothermal is the smallest reusable source (0.4% of total US electricity generation) in the country's electricity portfolio.

A 2000 paper by Charles F. Kutscher indicated there potentially could be over 20,000 MW of geothermal power available in the western United States, mainly in Nevada, Utah, Idaho, and Oregon. But the paper also pointed out that

> while geothermal energy is very clean, it is not as renewable as solar and wind energy. Like the energy of the sun, the energy within the earth is immense and has a lifetime measured in billions of years. However, unlike the use of sunlight, tapping into local sources of the earth's heat can result in a temporary decrease in the local amount of energy available. Reinjecting geothermal fluid that remains after steam is extracted helps preserve the fluid volume of the reservoir. However, even with reinjection, the heat content of the reservoir gradually declines.[54]

The author estimated that if overused, some thermal reservoirs could take hundreds of years to recover. Again, this is an example of a thermal balance, where the amount of heat pulled out to make electricity must be balanced by the amount of heat flowing in from lower layers of rock. Based on this, we cannot assume we can achieve the full 20,000 MW.

Another paper published in the *Oil & Gas Journal* in 2005, laid out the possibility of using old oil and natural gas wells in Texas and other regions for geothermal power.[55] The paper indicated that several oil/gas wells had encountered high-temperature reservoirs at over 250 °F at depths of up to 15,000 feet below the surface. Developments in this area should be encouraged and hopefully, some electric generation is possible in old oil field regions of Texas, New Mexico, Louisiana, Pennsylvania, and the Dakotas.

For this discussion, let's assume another 5,000 MW of geothermal can be achieved in the next thirty years. That may seem conservative, but policy needs to work with realistic estimates rather than hopeful but unlikely numbers. That 5,000 MW of geothermal would provide an additional thirty TWh to the US electric supply.

A recent development is the geothermal heating and cooling of homes, which is also referred to as Ground Source Heat Pumps. These systems take advantage of areas below ground that remain at a constant temperature throughout the year. In some cases, systems are installed under ponds where the temperature is constant, usually 60–75 °F. Water is pumped from the house to the constant temperature area and returned to the house. In the summer, heat is removed from the house and "pushed" into the underground area, and in the winter, heat is "pulled" from the underground area to heat the house. The installation costs are several times more than a standard heating/cooling system. The advantage is that geothermal heating/cooling systems are much more efficient and pay for themselves over time by using less energy and having a longer design life than furnaces or air conditioning units.

NUCLEAR

Nuclear power plants represent one of the cleanest and most controversial power sources. The recent nuclear tragedies of Chernobyl (1986) and Fukushima (2011) are the clear examples of the risks, where nuclear meltdowns released large amounts of radioactive material damaging the environment and requiring evacuations and exclusion zones surrounding the accident. The United States experienced a partial meltdown of the Three Mile Island facility in Pennsylvania in 1979. At present, the United States generates 809.4 TWh of electricity from ninety-six reactors that have a combined 98,119 MW of installed electric generation capacity. These nuclear plants run year-round and have an efficiency of 94%. But nuclear plants are some of the most expensive electrical power-generating installations and have been plagued with cost overruns, schedule delays, and protests from environmental groups and local residents. While nuclear plants are carbon neutral (no CO_2 emissions), the plants produce a lot of heat. Of the heat generated by the nuclear reactions, 66% goes into cooling systems, either into the air or into rivers/lakes if they are nearby. That heat can be recovered and supplied to residential, commercial, or industrial users for their processes, but at this time that isn't done.

There are new nuclear-reactor designs that have been proposed, and the Department of Energy is looking to build prototypes that use more advanced designs that are both safer and more compact. These new small modular reactor (SMR) designs might be available for commercial use in two to three years, but we need to keep in mind that construction of nuclear plants does not happen overnight. Previous projects have taken up to ten years to build, and a few have dragged on far longer. The nuclear industry thinks these new SMR plants will take only five years, but for policy purposes, it should be assumed that any decision made today will produce working nuclear plants in twelve to fifteen years.

New safer designs coupled with capturing and using waste heat from the nuclear reaction to support nearby industry and home heating could make nuclear a strong contender in the mix of future electricity sources. Those are the positives; the negatives are the risk of another meltdown, which can be minimized with safer technology, but there remains the fact that a nuclear reactor generates radioactive waste that would need to be managed.

OTHERS

Other forms of renewable energy, burning wood and waste, generates a small amount of electricity each year (~57 TWh, 1.4% of the total US electricity). While that has the potential for expansion, I am not going to address it at this time. The focus needs to be on the major sources of renewable and clean energy that can be expanded within the next twenty years.

With the Green New Deal, Democrats want to replace the fossil fuel that generates 2,581.7 TWh of electric power by 2050. But there are other areas we need to consider in which moving away from fossil fuels will increase the amount of renewable electricity that is going to be needed. Especially when you consider transportation where oil is king. US gasoline consumption was 142.71 billion gallons in 2019. Add the 47.2 billion gallons of diesel fuel consumed in 2019, and you have a large segment of the US transportation energy that must be replaced. All this fossil fuel energy was used to drive 3.24 trillion miles in the year 2019.[56] Electric cars average between one and six miles per kilowatt-hour (kWh), for now assume three miles/kWh. This is ignoring that about 10% of the miles are by large trucks, but just for this discussion, assume three miles. That would require an additional trillion-kilowatt hours of electricity or 1,080 TWh.

Another consideration is fossil fuel used in homes and commercial businesses. According to the US Energy Information Administration, residential homes use natural gas for heating (59.5 million homes) and cooking (39 million homes). Fuel oil/kerosene is still used for home heating in 6.9 million homes, mainly in the Northeast. That is about 6.1 quads of energy. Now fossil fuel heating and cooking isn't as efficient as electrical, as some of the heat goes up the chimney and out into the atmosphere, so let's assume we need to replace about 2.5 quads of energy for residential use. About 3.4 BTUs is equal to a watt-hour, so 2.5 quadrillion BTUs converts into about 733 TWh.

Commercial locations (businesses, office towers, etc.) use about 4.6 quads of fossil fuels for heating and other purposes. For this discussion, assume two quads of electric energy is required to replace the fossil fuels. That's another 590 TWh that is needed.

Industrial applications use about 31.2 quads of fossil fuels. Some of these are applications in which plastics and other items are being made, and each of these areas will need to be addressed individually to determine the best solution for each issue.

So, the question is how to add approximately 4.9 trillion kilowatt hours of electricity to America's electrical system without bankrupting the American government or stressing society with electrical shortages or massive change in a short period. Unintended consequences also need to be considered and hopefully avoided. Some people will point out that this assumes no growth in demand, which is true, but US electricity demand and overall energy demand has remained constant for the last twenty years. While population increases have occurred, improvements in efficiency have caused the energy usage per person to decline.

That adds up to 4,900 TWh of new renewable or nuclear electricity needed by 2050. Assuming that additions of wind and solar will continue and that the potential for hydroelectric and geothermal can be achieved, over the next thirty years, that adds 1,300 TWh of power per year by 2050. Where will the remaining 3,600 TWh come from?

Another factor to consider is that US electricity demand is not constant. The energy demand for electricity changes throughout the day and the seasons, with summertime highs in July of 650,000 MW instantaneous demand during the day and lows at night of 400,000 MW instantaneous demand. Contrast that with April's highs of 450,000 MW instantaneous demand during the day and lows of 350,000 MW instantaneous demand. This changing demand for electricity is one of the major reasons why electric power generation must be flexible. Due to the intermittent nature of wind and solar, a massive increase in wind and solar is going to result in a massive need for energy storage systems like pumped hydro or batteries.

The Democrats are talking about an epic effort to replace 80% of the electrical infrastructure and almost 95% of the transportation infrastructure in the next thirty years. And they seem to expect the government to do this.

21.
A Realistic Green Deal

The law of conservation of energy tells us we can't get something for nothing, but we refuse to believe it.

—Isaac Asimov

The fact that the United States has an almost constant need for ~300 million kWh each hour of the year shows that some form of constant electricity supply will be needed. And it will need to be much greater than 300 million kWh with converting all the fossil fuel energy from heating and transportation into a "green" source. The required power storage to accommodate wind, solar, and hydroelectric would be incredibly large to match the changing energy supply with changing energy demand. Solar works *almost* in sync with swings in power demand during the day and the year. Wind is usually out of step with demand but is more efficient. Nuclear and geothermal plants work best as base load, running continuously at maximum output to achieve maximum efficiency, but geothermal has limited growth potential, and nuclear comes with all its long-term concerns and risks. One energy source is not going to solve the problem. Decisions need to consider a mix of new energy sources and storage options.

The reality of selecting that mix is that the government has not been a wise chooser of winners and losers in the economy or society. When the government gets involved with incentives, tax refunds, or direct control of projects, it has not had a good track record. Not only our government, but worldwide, all governments have failed at micromanaging. Look at Communist countries where five-year plans became disasters. Capitalism has worked because when the economics and incentives are understood, capitalism tends to make the right choices. The concern must be that people have invariably found a way to take advantage of incentives and refunds without providing the envisioned benefit. The cleanest and simplest function of government is to tax the things we don't want and then spend taxes on the services that the government is best able to supply—defense, common infrastructure, and laws to protect society. Innovation these days rarely comes from government. The solution is to harness America's creativity and industriousness to solve this. Not only that, but the best solutions are simple, so what is a simple government solution?

There is a lot of discussion about green incentives, carbon taxes, and carbon markets. Carbon impacts are debatable, and carbon markets/credits invite people to find ways to "game" the markets, as we see with other forms of government intervention. The easiest solution is a tax that everyone can understand—a tax that is increased slowly over time that will encourage American society and capitalism to reduce our fossil fuel usage. The initial proposal is a tax per BTU paid by the energy producer or importer on all fossil fuels. The BTU content of natural gas, oil, and coal is easily determined and confirmed, and people can't argue with it. The energy companies already use BTUs in transactions among themselves to determine how good or bad a batch of fossil fuel is. The tax would start at $0.10 per 80,000 BTUs and increase $0.05 per year for eighteen years until it was $1 per 80,000 BTUs. So, every coal-mining company or oil/gas company that produces or imports fossil fuels would pay the tax and pass along those costs to the consumers. That means that as power plants burn fuel to generate electricity, they would pay the tax indirectly and as such the fossil fuel plants become more expensive making non-fossil fuel electric sources more economical. And finally, it means that anyone filling up their vehicle is paying the tax as well. No exceptions, no special carve-outs for specific industries, no exemption for any exports of fossil fuel from the United States. Anyone using fossil fuel pays the tax, from small restaurants with coal-fired pizza ovens to giant industrial plants, it is the only way to be fair and equitable in this effort. In the first year, this tax would generate about $100 billion, but after it fully ramps up, it would generate a trillion dollars, a strong incentive for people to reduce their hydrocarbon usage.

Progressives are likely to be appalled by this; they will see it as a regressive tax that affects the middle class and poor. The problem is that almost 90% of Americans make less than $200,000 per year and fall into that definition of middle class or poor. To make the change to greener energy is going to require all Americans to change their habits, not only with "greener" sources but also being more energy efficient. Not providing an incentive to change from fossil fuels is going to seriously hamper that effort. And by doing so as a tax, it maintains American freedoms and choices. Want to own and drive a 1980s muscle car? Go right ahead, but you're going to pay more to fill it up with gasoline. Don't want to insulate the attic or install a more efficient air conditioning unit? Sure, but you're going to pay more.

There appears to be an effort on the progressives' part to somehow steer America toward a "green" future without using taxes, as it is the dreaded *T* word that no politician can utter except when applied to the wealthy. But this gets back to the problem that incentives and carbon credits are indirectly addressing the issue and can be "gamed" by people. How did we address smoking? In 2009, we raised the federal tax from $0.39 to $1.00 per twenty-cigarette pack. States add more tax ranging from Virginia at $0.30 per pack to New York, Connecticut, Rhode Island, and Washington DC at over $4 per pack. Studies have shown that raising the price on cigarettes directly reduced the amount of smoking in the United States.

But this fossil fuel tax needs to be taken a step further. Imports of foreign goods would arrive in this country having been generated with fossil fuels and transported across the ocean using fossil fuels, so there needs to be a BTU import tax. This would encourage foreign countries to also reduce their fossil fuel use and to not penalize American products whose cost is impacted by the BTU energy tax. It would form two parts. How many BTUs did it take to make the product, and how many did it take to transport the product to the US border? The easiest solution is to arrive at an average BTUs/pound for manufacturing and transport and let the weight of the product determine its tax. This BTU import tax would be on all products, even those that are tariff-free due to trade treaties. When a shipping container arrives, weigh it, and charge the shipper a tax based on the weight. The container doesn't leave the port until the tax is paid. Unfair? Too bad. To keep this a simple solution, there's going to be some unfairness, but better a little unfair than overly complicated.

There is one last issue to be addressed, and that is importing electricity. There is the potential that people could build fossil fuel power plants in Mexico or Canada and import electricity into the United States, avoiding the BTU tax. For that reason, any imported electricity should be taxed at $0.01 per kWh for the first year, rising to $0.10 per kWh (ten kWh is generated on average by 80,000 BTUs), regardless of whether it can be shown that the electricity is 100% renewable or nuclear. Why? Because a green electron is the same as a fossil fuel–generated one, and no one can really say it is green. Unless the country has gone to a 100% non-fossil grid, its electricity gets taxed if it is imported into the United States.

The taxes generated would be used in the following ways

- 50% to pay down the national debt, not pay the interest, but pay down the principal of the national debt
- 10% to fund research and development of better energy sources and energy storage (new or better forms of nuclear power plants, hydroelectric, geothermal plants, wind turbines and solar panels)
- 40% for construction of infrastructure (energy storage, efficient power lines, and mass transit)

For infrastructure, the government would lend the money to developers of the infrastructure, each of whom would have to submit a proposal so that the most cost-effective would be awarded first. But the projects would also be evaluated after they began operation, and those that actually met their cost, schedule, and benefit to society would have 50% of the loan forgiven. This may seem strange, but it would encourage proposals to be accurate and beneficial. Too often, government infrastructure projects that are funded by the government suffer schedule delays and cost overruns that exceed what industrial projects financed by banks deal with. This would move the government into a banking-style investment and would reward projects that benefit society but keep them as cost effective and financially responsible solutions.

Infrastructure does not include electric power generators but does include long-duration energy storage. Any electric power plant needs to be able to meet the economic drivers that are present in the market. Coal power plants are dying out due to economic forces and should not be propped up, nor should "green" power plants have the economics tilted in their favor. Capitalism and economic forces do a good job of picking winners and losers over the long term. The addition of the BTU tax to fossil fuel plants will help by making renewable energy and nuclear more economical.

Long-duration energy storage is included because studies have shown it usually cannot be justified by electricity prices alone. While the Department of Energy has established their "Energy Storage Grand Challenge"[57] with a goal of $50/MWh for storage costs, none of the present technologies, even pumped storage which is the bulk of existing systems, can afford to build new facilities without tax credits and other incentives. Foreign countries, especially China with 50,000 MW of pumped storage in construction, are building long duration energy storage, but the only way they can do it economically is with government funding. Energy storage will become critically important as wind and solar produce more of the electricity for the country so the best time to build storage is now to support that future growth. Studies are showing that countries with high levels of wind and solar, but low levels of energy storage have higher electricity costs than countries with the same number of renewables, but higher levels of energy storage.

Great Britain and Germany both use over 10% wind and solar, but only about 3% pumped storage compared with their electricity grid. The electricity costs that consumers pay in these two countries are over twice as high as Austria's which also has 10% wind and solar but has enough pumped storage as 18% of their electric grid, thereby allowing them to deal with the variability of the wind and solar.

Likewise, efficient transmission lines need to be part of the infrastructure development. The ability to move wind from the Great Plains and solar from the desert Southwest will become more critical in the future. An example is the ERCOT market in Texas. Massive wind farms have been built in the northern panhandle region and in the southern corner along the Gulf Coast. On windy days, excess wind energy is limited by transmission bottlenecks and can't reach the areas where it is needed resulting in price fluctuation across Texas. Large price differences between the sources of wind energy and Texas cities where it is needed are not uncommon due to these transmission line bottlenecks.

The tax proposed would begin with motorists paying about $0.15 per gallon of gasoline and then rising to about $1.50 per gallon at the end of the eighteen-year period. The motorist wouldn't pay the tax directly, since it would be paid by the entity that produces or imports the original fossil fuel, but they will feel them all the same. European taxes on a gallon of gasoline are $2.00 to $3.00 per gallon. The United States at the present has a federal tax of $0.184 per gallon with states tacking on their own tax. California is the highest with a state tax of $0.612 per gallon for a combined tax of $0.796 per gallon.

Conservatives may question why we are bringing our taxes in line with Europe. The point is not to copy Europe; it is to change American behavior to reduce fossil fuels. Compared with other developed countries, American fuel taxes are low and provide no incentive to change energy habits. The federal tax on gasoline was raised to $0.184 per gallon in 1993, and it hasn't moved in the last thirty years. An additional tax of $0.15 per gallon rising to $1.50 per gallon over eighteen years is not excessive, and it is something that not only the country can tolerate, but that it needs to help transition away from fossil fuels.

And progressives may think that this is not enough. The Green New Deal demanded high-paying union jobs in clean industries, as well as carbon-emission requirements. Any governmental attempt to regulate industries to such a degree to guarantee high-paying jobs or any other requirements is going to involve more bureaucracy with the potential for unintended consequences and not achieving the real objective—to harness American capitalism to solve this problem, since it was American capitalism that created it. Increasing the costs of fossil fuels encourages Americans to change their behavior, drive less, take mass transit, and use less fossil fuel. The revenues can then be used to build infrastructure, encourage new technologies, and pay down the national debt without a massive expansion in federal bureaucracy, which is likely to be inefficient, both in terms of cost and energy.

The Energy Policy

- Implement a tax paid by the producer or importer of hydrocarbon that would start at $0.10 per 80,000 BTUs and rise $0.05 per year for 18 years until it was $1 per 80,000 BTUs.
- An additional tax would be levied on all imported goods based on their weight to reflect the BTUs required to make and transport.
- Funds generated would be dedicated to pay down the National Debt (50%), to fund research and development of better energy sources (10%), and for construction of energy storage, efficient power lines, and infrastructure (40%).

22.
The Fate of Oil

You cannot escape the responsibility of tomorrow by evading it today.

—Abraham Lincoln

If at first you don't succeed, try, try again. Then quit. There's no point in being a damn fool about it.

—W. C. Fields

Conservatives and oil executives will look at the fact that US oil production has reached historically high levels and wonder why. Why turn away from oil and natural gas now and risk the American economy? This may seem like the absolute height of folly to them. Especially if they feel it's unlikely that any other nation will copy our effort, and we are only giving foreign nations a competitive edge. Disregarding climate change, some will see no reason to do anything differently, no reason to walk away from hydrocarbon and the cheap energy that it provides to power an economy.

But there is a reason.

For the better part of my over thirty years as an engineer, I designed and built projects for the oil industry. A number were massive offshore oil platforms that achieved their maximum production rate in the first year or two. Whenever I returned to them after those first years, I always found facilities producing far below that maximum rate. It is the very nature of an oil field that its best years are at the beginning when the oil starts to flow, but then depletion kicks in, and the oil field produces a little less every day. Until ten to thirty years later, when the oil field operator is trying to decide how to shut down operations, remove the platform, and abandon the field.

This hasn't been an issue in the past since there has always been another oil field to move on to—a new project in a new location.

But consider the history of oil and gas development in the Gulf of Mexico off the southern United States. The first well was in 1937 and it was drilled from a wooden trestle in Texas extending into the gulf at a water depth of thirteen feet. The first truly offshore wells in one hundred feet of water weren't drilled until 1955. A water depth of 200 feet was broken in 1962, and the record of 1,000 feet was shattered in 1979. Since 2014, more than 80% of the oil production from the Gulf of Mexico comes from wells in over 1,000 feet of water depth, what's called the Deepwater. The offshore oil industry is producing more from Deepwater than ever before because they have depleted most of the oil located in the shallow areas. This is a known fact. Oil fields are depleting, and this is forcing oil companies to look to even deeper water for the new fields that they require.

There have been multiple times in the past when the US oil industry has been in decline and the thought was that "peak oil" had occurred. Peak oil is the day when the United States produces the most oil it ever will, and from that point on, the oil industry will be in permanent decline. Each time people thought this had happened in the United States, the industry has had some breakthrough that has delayed peak oil: The opening of the Trans-Alaska Pipeline (1977), Deepwater Gulf of Mexico production (1990–95), and US Shale production (2008–15). Each breakthrough has turned production around and pushed back the time when peak oil would finally occur in the United States.

And while we can hope for another breakthrough just beyond the horizon, it is the nature of the horizontal shale fracking boom that makes me think horizontal fracking was the last rabbit the oil industry could pull out of its hat. Early oil discoveries came from reservoir rock, sedimentary deposits that were once ancient beaches or riverbeds with a high degree of porosity. Porosity is the volume of empty spaces within a rock and a measure of how much liquid a volume of rock can hold. Think of sandstone drink coasters that absorb the water that beads off the side of a cold glass. Underground, these formations can hold billions of barrels of fluid between the sand particles, and sometimes those barrels were oil. The early oil discoveries in Pennsylvania and Texas were the result of drilling into sedimentary deposits packed with crude oil.

It took a while for geologists and petroleum engineers to figure out that the reservoir rocks weren't the source of the oil, they were just where it accumulated and was stored. The source rock was shales and limestone deposits that were rich in organic compounds. And when those source rocks were placed under pressure from the layers of rock that had formed above them and built up over time, the organic material would change into natural gas and crude oil. In oil field terms, source rock under the right pressure and temperature to form crude oil were called "kitchens," and they began to look for kitchens near sedimentary rocks, hoping to find big reservoirs. But they also learned that you could drill into the shale kitchens and produce oil directly from them. The problem is the poor porosity of the shale rocks meant that shale oil wells produced very low flow rates and rarely recovered the cost of drilling the wells. It was far better to drill into the reservoir rocks where the higher porosity meant high flowrates that made the sedimentary rock oil wells money-making machines. That is, until the reservoir rocks had been drained of all recoverable oil.

The shale fracking boom was based on two discoveries: the ability to drill down and then bend the drill bit from the vertical to the horizontal and drill horizontally through the shale layer. As of this writing, the longest horizontal segment in the United States is now almost four miles long, and internationally even longer horizontal segments have been drilled. That may not seem like a massive breakthrough, but the amount of oil produced by the well is determined by how much of the oil-bearing layer is penetrated by the well. A simple vertical well and fifty-foot-thick shale layer only touches fifty feet of the oil-bearing shale. That same fifty-foot-thick shale layer with a 20,000-foot horizontal section in the shale layer touches 400 times as much shale.

Now add the fracking technology. Fracking has been around for a long time; most vertical wells are "fracked" in the oil industry to improve their production. Fracking is pressuring up the well bore to crack the oil-bearing rock and create fissures. Realize that it is the surface area of the exposed oil-bearing rock that the oil comes from. Making lots of cracks within the rock and then adding sand, known as proppant, to keep the fissures open creates more surface area and increases the amount of oil that will flow into the wellbore and up to the surface.

It was fracking combined with long, horizontal well bores that allowed shale wells to go from vertical wells that made little to no profit to horizontal wells that made a huge profit and allowed American oil production to increase from 1.8 billion barrels in 2008 to 4.5 billion in 2019. The problem is we are drilling out the kitchens, and once those are produced, where will new oil come from in the United States? There doesn't appear to be any.

A further problem is that those horizontal shale wells also have very steep production decline curves. Every existing well has a production profile that shows how production decreases day by day from the individual well. The well might produce 1,000 barrels per day for the first week but come back in a year and production will be a fraction of that. The EIA publishes a drilling productivity report for US shale regions. The December 2022 report showed over 350 drilling rigs in the Permian region of Texas/New Mexico that would produce new wells with a combined production of 370,000 barrels/day. But the report also showed that the existing wells in the Permian had declined by 333,000 barrels/day, meaning that all the effort of those 350 drilling rigs would result in Permian production increasing by 37,000 barrels per day, a 0.6% increase from the previous month. Nationwide, 674 drilling rigs working the seven major shale fields resulted in an overall increase of 94,000 barrels/day of oil and 535 million cubic feet/day of natural gas. About 86.5% of the rigs working in our seven major shale fields are there just to replace lost production. And that's assuming you have enough of what are called Tier 1 wells to drill. These are the good shale wells that everyone wants to drill, and when you run out of those, you need to drill the Tier 2 wells that aren't as good, and you will need more wells to make up the lost production. And Tier 2 wells not only provide less initial production, but they also decline faster than Tier 1 wells, so your next round of replacement will need to drill even more due to the Tier 2 well's higher decline rates. And don't ask what happens when you run out of Tier 2 wells to drill; you don't want to see what the numbers are for Tier 3 or Tier 4.

This analysis was confirmed by a recent *Wall Street Journal* article.[58] The article pointed out that the best wells drilled in the Permian for 2022 were producing 15% less than the best wells drilled in 2017. The implication was that the best acreage has been drilled, and that it is possible that production from the Permian would plateau, then begin to decline. This has major implications; the immediate result is decreasing production resulting in higher pricing necessary to spur shale companies into increased drilling. It also foreshadows the United States returning to higher levels of oil imports making us more vulnerable to international oil/gas market impacts.

There is still a lot of oil in the United States and worldwide, but it won't last forever, and at some point, world oil production will permanently decline. My fear is that trying to keep production rates increasing and meeting society's demand will push the oil industry to produce at ever higher rates until a precipice will be reached, and then the decline in oil production won't be driven by a declining demand, but a rapid decline due to lack of new oil fields. Such a rapid decline would be disastrous for society worldwide since there wouldn't be time to develop alternatives. So, there is a good reason to develop alternatives now. Especially when the United States has only 3.2% of the world's oil reserves, that is, the known proven reserves of oil that haven't been produced yet. And the volume of proven future oil in the United States would last less than twelve years at present rates of US oil consumption. Of the world's oil reserves, 85% is either in countries hostile to the United States or located in the Middle East, which are vulnerable to disruption. Further, with all the proven oil reserves in the world, if you assume the present rate of consumption without considering an increase in consumption, there is only forty-three years of future oil supply in the entire world.

When people speak about oil reserves, they use different numbers that refer to three types of reserves defined by the Securities and Exchange Commission: proven, probable, and possible. Proven reserves, which are sometimes referred to as 1P, are estimates of future oil production that have a 90% chance of production. Probable reserves means production that has a 50% chance of occurring. And finally possible reserves refers to production that has a 10% or less chance of occurring. People against switching from fossil fuels to renewables are quick to bring out the 3P (proven+probable+possible) reserve numbers and using them to justify continuing with a "drill, baby, drill" attitude. But think about what that means: they will say we have "hundreds of years of fossil fuels" when they are basing it on oil reserves that have a 10% chance of occurring. Having watched oil companies write down their 1P reserves because they were overestimated, the idea of betting our children's future on a 10% chance seems pretty unrealistic.

Now, we are always discovering new oil reserves, and the oil companies report their *reserve-replacement ratio* (RRR). A 100% reserve-replacement ratio means a company is finding enough new oil to match the amount they are producing. The cumulative average RRR for the major oil companies from 2016 through 2020 was less than 100%. ExxonMobil led the pack with a cumulative RRR of 89%. BP was 48%, Shell 29%, Chevron 16%, and ConocoPhillips 78%.[59] The report from Rystad Energy also lists the Italian oil company Eni and the French oil company Total. The companies listed in the report produce only 8% of the world's oil, but they are required to report their RRR, where the national oil companies like Saudi Aramco are not. People may wonder why the big oil companies, including Saudi Aramco, are embracing solar and wind. The reason is that with the entire oil industry reporting reserves-replacement ratios far less than 100%, they can see an end to the age of oil.

That is why even if climate change is not considered, developing alternative energy sources to fossil fuels should be viewed as important and necessary. When climate change is considered, with data in United States showing that temperature trends are increasing, then the requirement for change becomes critical.

The oil industry needs to begin to look at changing. The future is no longer as bright as it was ten years ago, but there are alternatives. A consideration for the oil industry is the need for energy storage. Energy storage will become a major need in the future. While everyone at present is investing in grid-level battery systems, the demand for batteries will skyrocket as cars transition to electric, and we could see shortages in lithium and other rare earth metals that batteries require. The limited life of battery systems also works against them.

There has been some work with compressed-air storage. During times of excess electricity, the excess is used to compress air and store it underground and then when demand is high, the compressed air would generate electricity. While the compressed air system has a low efficiency (electricity returned from storage versus electricity put into storage of only 40–52%), it has a longer design life than batteries and doesn't need rare earth minerals. Old natural gas reservoirs could be used for this.

A twist on the idea of pumped hydro storage is that old oil and natural gas reservoirs could be used for storage in a pumped hydro system. The reservoir would have to be partially gas-filled at a high enough pressure to have a positive pressure at the top of the wellhead. Pumping water into it would increase the pressure in the reservoir, and that would allow the water to be used to drive a turbine later to make electricity. Not sure how this would stack up against the existing pumped hydro using lakes in the mountains. There is limited expansion for existing surface-pumped hydro designs, so these underground reservoirs could be in high demand for electric storage and not only make this feasible but generate a high revenue for it.

Another thing for the oil industry to consider is that the knowledge of drilling is something that the geothermal industry needs in order to increase geothermal power. There have also been studies that looked at existing oil and gas wells that are in high-temperature reservoirs as possible sources of geothermal energy.

The point is that rather than fight the trend toward less hydrocarbon energy, it may be time for the oil and gas industry to consider what it can do to help that trend.

Now, this does not mean that the oil industry is grinding to a screeching halt. There will be a need for oil for a long time, though it will be in decline, and one thing that can be done to help the oil industry and the American economy is to establish an oil bank. The United States has the Strategic Petroleum Reserve (SPR). The SPR has the capacity to store 714 million barrels of oil. The SPR was established to reduce the impact of disruptions in supplies of petroleum. The idea is to dedicate half the storage volume in the SPR as an oil bank that will sell oil when prices are higher, say $20 to $30 over the yearly average price for a barrel of oil. And it would buy oil when prices are lower, say $20 to $30 under the yearly average price for a barrel of oil. The result is that the oil bank becomes a stabilizing influence on both US and world oil prices. The fluctuation of oil prices has been a major problem in recent years for the oil industry and the economy.

In the past, OPEC and Saudi Arabia have worked to stabilize oil prices. But in 2014, with US oil stockpiles starting to climb and oil prices starting to fall, OPEC and the Saudis were unwilling to cut back production until prices had reached rock bottom and the entire industry was hurting. Going forward, we cannot count on Saudi Arabia or OPEC to help stabilize oil prices, a new mechanism is needed.

Not only does the oil bank stabilize world oil prices, but it also has the potential to make money. Buying cheap and selling high can produce a cash profit, and while most of it will be eaten away by maintaining the oil storage, it makes this a very cheap investment.

THE OIL BANK POLICY

- Convert 350 million barrels of oil storage in the Strategic Petroleum Reserve into the US Oil Bank.
- The mandate of the US Oil Bank would be to set prices when to buy oil when the oil market is oversupplied, and prices are below average and to sell oil when the market is undersupplied, and prices are above average. It would then sell and buy oil as required by the market and price targets set.

23.
The American Transportation Policy

A good science fiction story should be able to predict not the automobile but the traffic jam.

—Frederik Pohl

In 2018, Americans drove 2,917,159,000,000 miles in cars, motorcycles, and light-duty trucks, essentially in their personal vehicles. The number of miles that transit buses and trains drove was a fraction of that: 4,168,000,000 miles, less than 0.13%. To reduce America's use of fossil fuels for transportation, two kinds of action are needed. First, converting fossil fueled cars, light trucks, and motorcycles to non-hydrocarbon vehicles, and second, by having more Americans use public transit or walk, will reduce the number of vehicles to be converted. So, what can be done for each vehicle type?

The Bureau of Transportation (ww.bts.gov) lists over 250 million "light duty vehicles," a category that covers sedans, coupes, small pickup trucks, and sport/utility vehicles. Of that total, over 2 million are electric cars. The entire US car sales per year, including fossil fueled, hybrids, and electrics, totaled 13.7 million for 2022. Only 762,883 were electric (5.1% of all sales). The EV industry is celebrating as it was a massive increase compared with 2021, when electric vehicle sales were 462,247. Electric vehicle sales are expected to reach 7% of all vehicles in 2023. To covert the remaining 247 million American cars to electric will require a major effort. It will require all of the automotive industries to transform from internal combustion engines to electric motors and batteries. The retooling and changeover of all the factories will require a herculean effort, especially considering a thirty-year target to replace most cars. But there may be a problem. We may face a shortage and competition for lithium for the batteries. Worldwide lithium production was 130,000 tons in 2022, a new peak in production which will be exceeded by 2023 when final numbers are in. It was 95,000 tons in 2018 but dropped to 77,000 tons in 2019 due to excess production and low prices for lithium. The bulk of the lithium comes from three countries—Australia (55,000 tons/yr.), Chile (26,000 tons/yr.), and China (14,000 tons/yr.). China is also the largest processor of raw lithium ore with most of Australia's production going to China to be turned into batteries.

A typical electric car has about twenty-two pounds of lithium, and most cars are small, so we need to assume that with light trucks we are going to need more. To replace the 247 million fossil fuel cars and trucks in America, we are going to need about 2.5 million tons of lithium, not to mention lots of other stuff. And that is not considering all the other needs for lithium or the other countries that also need lithium for their electric cars and batteries. A mammoth increase in the production of lithium could have environmental consequences as well as global political consequences, as the major sources are not in the United States. And lithium isn't the only issue. Copper will also become a critical material for electric cars as well as several others. So, that brings up the question of what other options are there?

HYDROGEN

Hydrogen is the darling of the Department of Energy. While there have been discoveries of naturally occurring hydrogen, hydrogen production from drilling is in its infancy and the only commercial development has been in Mali. So, until a hydrogen drilling revolution occurs, we can assume that hydrogen isn't an energy source by itself. The bulk of hydrogen can only be found naturally as part of a larger molecule. Water is two hydrogen atoms with an oxygen atom. A major part of natural gas is methane, which has four hydrogen atoms bonded to a single carbon one. So, until drilling for hydrogen becomes a reality, and I don't think it will, you need to make it through chemical or electrical reactions with existing molecules, water and methane being the best two. Natural gas can be converted to hydrogen by removing carbon from the molecule. The problem is, what do you do with the carbon atom, especially as we are trying to reduce the amount of carbon in the atmosphere? There are several ways in which natural gas and other fossil fuels can be converted to hydrogen, and each has pluses and minuses, but all only get hydrogen that represents 50% or less of the energy that the fossil fuel started with and all struggle with what to do with the carbon atoms generated.

Electricity can be used to convert two molecules of water into two molecules of H_2 (individual hydrogen atoms are unstable and combine into a semi-stable molecule that has two hydrogen atoms) and one molecule of O_2 (same thing, oxygen atoms like to hang out in pairs) through a process called electrolysis. This process is about 70 to 80% efficient, meaning that of the electricity supplied, the resulting hydrogen has 70–80% of the energy stored in the hydrogen. The hydrogen can then be used either in a fuel-cell-powered car where it converts to electricity via a fuel cell, much like the ones used in NASA's Apollo missions, or in a hydrogen internal-combustion engine, which is similar to a normal gasoline-powered engine.

Now for the problems. While the process of making hydrogen is decently efficient, hydrogen is made as a gas at low or atmospheric pressure, then it has to be compressed from a large volume of gas at low pressure to a high-pressure gas at low volume for easier storage. That uses additional energy, making the process less efficient. Once you have high-pressure gas, then you need to have high-pressure vessels to store it in that need to be incorporated into refueling stations and cars. Then the cars need either hydrogen fuel cells, which are completely different from internal-combustion engines, or hydrogen internal combustion engines, which burn at such a high temperature that all the internal metals and seals within the existing engines need to be replaced. The result is a transportation system that can't use any of the parts of the existing diesel/gasoline system and relies on a high-pressure gas refueling system and hydrogen engines with an overall efficiency that is fairly poor, worse than the gasoline and diesel they were out to replace. And oh, by the way, hydrogen is a very explosive and flammable gas. Have you heard about the German passenger airship the *Hindenburg*, a giant hydrogen-filled dirigible that exploded in flames and crashed in 1937 while trying to land in New Jersey?

So, we now have hydrogen in a high-pressure form that makes it about 30–40% efficient and we can use it in cars, home heating, or electricity generation, respecting that like natural gas it can be explosive. The problem is that it needs an entirely new infrastructure. In engineering, there is a phenomenon called hydrogen embrittlement, which affects most regular steels. Embrittlement means that the hydrogen atoms have made the steel more brittle, and the steel can crack and fail under stress. To safely transport and store hydrogen requires special stainless steels or composite structures that can resist this embrittlement, none of which exist in the present infrastructure to transport natural gas, so you just can't start flowing hydrogen in old natural gas lines.

Hydrogen represents a way to convert electricity into an energy storage medium, but the problem is that it has the highest values of energy by weight, but some of the lowest when looked at for energy by volume even when compressed. Gasoline has an energy density of 46.4 megajoules per kg whereas hydrogen gas at atmospheric pressure has an energy density of 125 megajoules per kg. So, by weight, hydrogen wins, but weight isn't the key issue here, it is volume. Gasoline is 34.2 megajoules per liter; hydrogen at atmospheric pressure is only 0.01. Even when you compress hydrogen gas to 10,000 psig pressure, its energy volume is only five megajoules per liter.

What that means is if you replace a gasoline fuel tank with a high-pressure tank with hydrogen, you get less range between refueling. Using a fuel cell instead of a hydrogen-burning engine improves the range, but not to the point that a gasoline/hybrid vehicle gets. Hydrogen cars using pressurized hydrogen for fuel have about 75% of the range of their gasoline equivalents and had to give up some trunk space to do it.

The reason the hydrogen economy hasn't taken off yet is that it needs a completely new infrastructure and it's less efficient from an energy volume standpoint. That doesn't mean it won't happen, just that at this point, it is hard to tell if this will be the long-term solution that so many people are hoping it will be.

ETHANOL

Ethanol is ethyl alcohol made from agricultural crops. Of the gasoline sold in the United States, 98% has some ethanol in it, typically E10, which is 10% ethanol and 90% gasoline. The reason for adding ethanol is not only to use a more "renewable" fuel, but ethanol helps the gasoline burn more cleanly and produce less pollution. Prior to this, methyl tertiary-butyl ether (MTBE) was added to gasoline to make it burn cleaner, but because of concerns over groundwater contamination and water quality, MTBE is now banned in most states and ethanol has taken over. Many modern cars are termed as flex-fuel and can run on fuel with up to 85% ethanol, but there are some drawbacks. Ethanol has a lower BTU content than gasoline—remember BTU content is the measure of how much heat is generated when the fuel is burned, so fuel that is 85% ethanol generates fewer miles per gallon than gasoline without ethanol, so the car doesn't go as far before it needs to be refueled. Also, the higher the ethanol content, the harder it is for older, non-flex fuel vehicles to use.

Now, the thought may be that this fuel is still generating heat and carbon dioxide, but it is working more like a closed loop. The growth of the agricultural crop traps the solar heat and uses carbon dioxide from the atmosphere to grow the crop. In terms of climate change, ethanol and other biofuels are carbon neutral. That idea is being debated vigorously by supporters and opponents of ethanol. Growing corn and converting it to ethanol requires fertilizer, tractors, transportation, fermentation, and distillation to get the ethanol output. Some argue that this is supplied by fossil fuels, but they could be converted to biofuels or electricity to achieve a greener footprint for the ethanol.

These points are still in contention, but there is one issue that is clear. The energy required to grow and process the crop to generate the fuel and then the efficiency of the internal-combustion engines means that the actual useful work being done by the internal combustion engine is a small fraction of the solar energy applied to the land where the crop was grown. So, converting large amounts of America's gasoline supply to ethanol is going to require a massive increase in land use. The reduction in availability of fossil fuels is going to negatively impact American farms to produce as much crop per acre as they do at present, so this runs counter to the need for more land for ethanol. The idea of how much land to dedicate to ethanol is being argued by the proponents and opponents of biofuels, but this misses the point. America is going to need something to help the transition from fossil fuels to a green transportation network.

A greener transportation network is definitely in the future, but the actual shape of the network is unknown. Many are predicting and prophesizing the details, but there is really no way to know the details of the future transportation network because there are so many variables to consider. And we should put a lot more thought into agriculture, long-distance trucking, and aviation. Electric vehicles with batteries are going to have a hard time filling these roles. That isn't to say it is a sure thing that these three roles won't morph into something completely unforeseen and won't need biofuels. But, in the present outlook, it is very likely that biofuels, specifically ethanol, will be needed to support at least agricultural machinery, long-distance trucking, and airlines, to help the transportation network make the transition.

A developed country isn't a place where the poor have cars. It's where the rich use public transportation.

—GUSTAVO PETRO

The last point to consider is mass transit. Mass transit, short-range within cities and suburbs and long-distance between metro areas, is going to be needed. The reason is basic efficiency. One of the core issues is that even with converting personal cars to renewable energy, there is a basic inefficiency of the individual personal car. When a person drives their car, they are using on average a 3,000-to-4,000-pound vehicle to move an average 180-pound American around town. Even when an electric car is being used, the personal car is far more energy expensive than a metrobus, subway, or train. This is because when driving a personal car, the 180-pound person needs the energy required to move 4,000 pounds of weight of the car and the person, while a train shares the weight of the train with all the passengers and the weight per person that you need to spend energy to move is less. Like using a dolly to move multiple boxes at once as opposed to having to carry them individually, the dolly needs less energy. Same thing for mass transit. Moving lots of people at once instead of by individual cars is much more efficient. This becomes even more apparent when you factor in the inefficiency of hydrogen or ethanol, which will likely be needed to allow personal vehicles to continue to be used in the future.

This is not an appeal to convert all Americans to mass transit. The suburban sprawl of most American cities does not lend itself to an efficient mass transit system. But there will need to be a change in how some Americans move around. And again, picking a winner right now is not the right decision. Electric buses and light rail within metro areas will be likely solutions in the immediate future. But the government needs to promote the idea and be ready to use some of the infrastructure funds generated by the BTU tax laid out in the chapter on the Realistic Green Deal. The pressure that the BTU tax will exert will cause American cities and suburbs to begin to morph into layouts that are more beneficial to walking, cycling, and mass transit. This can be seen in some metropolitan redevelopments here and in Europe, where this change in attitude has already occurred.

I knew I was going to take the wrong train, so I left early.

—YOGI BERRA

Long distance travel between metro areas is dominated by personal transportation and airplanes. In 2019, Amtrak posted a $30 million loss, which for the federally supported passenger rail service was actually an improvement from the past few years. The estimate for 2022 is that Amtrak will lose $1 billion as it tries to expand after the difficulties of COVID. Prior to World War II, passenger train service was a major method for anyone to travel long distances across America. After World War II, personal cars and airlines reduced the demand for passenger trains, which resulted in Amtrak taking over all passenger service in 1971 when Union Pacific and most other private carriers stopped. Union Pacific, Norfolk Southern, and other train carriers all still operate, but they now focus only on cargo. Even now, the long-distance trains for Amtrak are a financial liability, accruing major losses.

But in Europe, Japan, South Korea, and China, long-distance rail is not only healthy but actually increasing. In Japan and South Korea, the high-speed rail industry is not subsidized and actually makes money. In China, the government subsidizes the rail industry by $130 billion. Europe subsidizes the rail industry by €73 billion ($86 billion), but these are small fractions of the overall GDP. For China it's 0.9% of the national GDP, and for Europe, it is 0.5% of the combined European GDP. The reason Europe and China both subsidize the rail industry is it reduces their dependence on oil and allows internal travel, which is important to their economies. The United States doesn't view rail travel as important and only provides a small subsidy to Amtrak. Both intra-city and local rail travel in America is very limited. So, what's the difference between those countries and America? Well, first gasoline prices are substantially higher there. Wikipedia listed gasoline prices around the world in US dollars, and the lowest-cost countries at the time were Libya ($0.12/gallon) and a number of the Middle East oil suppliers like Iran ($0.20/gallon) where the governments subsidize the gasoline prices to keep them low.[60] At the time, the United States was listed at $3.79/gallon. Europe averaged $7.76/gallon, Japan was at $4.50/gallon, and South Korea was $4.58/gallon. China was at $4.35/gallon, the lowest when compared to other successful rail operators, but we need to realize that China is growing, and one of the reasons they subsidize their rail system is to make it affordable and to address the lack of availability of cars and roadways.

Lack of high-speed railways in the United States is the second factor that has made rail transport less successful than in other countries. The United States lacks the train infrastructure because in the 1950s the country chose a different path.

President Eisenhower is usually given credit for building the Interstate Highway System by signing the Federal-Aid Highway Act of 1956. That neglects some of the groundwork laid by President Roosevelt. Roosevelt wanted to build a national transportation system to rectify the problems that had been encountered during the massive movements of men and material during World War II and to have more jobs waiting for the returning soldiers after the war. But Eisenhower, after seeing the German highway systems, which had allowed armies to be moved rapidly, became a strong advocate for the system of interstate highways that are now called the Eisenhower Interstate Highway System. The purpose of the system was not only to encourage commerce between the states, but to allow the movement of military personnel and equipment around the country. It has been a resounding success, putting the passenger trains almost completely out of business.

But that was then. As the cost of fossil fuels increases to encourage Americans to use less, it will generate an incentive for people to use mass transit more, both within metropolitan areas and rail transportation between metro areas.

Now the return of passenger service to American rails won't be without cost. Most existing tracks in America's railroad system are limited to 79 mph. Even back in the 1920s, trains travelled faster than that, many averaging 85 mph or faster. The issue today is the congestion of freight trains using those same rails. The volume of freight trains has increased tremendously since the 1920s, so fast passenger trains need to look elsewhere. And why focus on fast passenger trains? Americans are going to be reluctant to use a system that takes days to get them across America when airlines do it in hours. And there is no reason not to expect high-speed trains to be competitive because the successful long-distance trains in foreign countries achieve speeds much higher than 79 mph. The Spanish AVE train between Madrid and Barcelona reaches up to 220 mph and makes the trip in two and a half hours. It is actually taking ridership away from the airlines doing the same trip. The direct flight takes an hour and twenty-five minutes and costs about the same. Other train systems easily exceed 150 mph in Europe, Japan, South Korea, and China, but they run on rail systems specifically built for high speed. At present, the United States has only a few miles of track rated for high speed. Some of the track between New York and Boston allows speeds of 150 mph. The New York to Washington route has some areas that allow speeds of 135 mph from the Acela trains by Amtrak. The Acela train itself is capable of 200 mph but, lacking the track, it can't achieve those speeds in the United States.

The Federal Railroad Administration defines the quality (speed rating) of American railroad tracks. Class 9 rail track is capable of 220 mph. At present, none exists in the United States.

To achieve true high-speed rail, an entire new rail system will need to be created, independent of the freight rail system. And that isn't a bad thing. Increased fossil fuel costs will make freight rail more attractive creating higher demand for freight trains. Which is very good for the freight rails, which have already been enjoying a renaissance that started in the 1980s. The high-speed rail system could also carry freight at a higher cost than low-speed freight, but this would replace the freight capacity presently being handled by air freight, which will be impacted by the increase in fossil fuel costs.

A very promising technology is magnetically levitated trains (maglev), where magnets built into the rail system cause the train to float above the rails on magnetic fields. This removes the friction losses that occur between the rails and the steel wheels of existing systems. Maglev was developed in the 1960s and '70s, and the first systems appeared in the 1980s. An existing maglev system between Shanghai airport and the financial district opened in 2003 in China and operates at 270 miles per hour. There are several low-speed systems operating at airports and over short distances in South Korea, China, and Japan. China and Japan are both building major long-distance high-speed systems. The energy efficiency of maglev trains is the absolute best of all the long-distance transportation solutions because they lack the wheel-to-track friction that every other system needs to overcome. Trains are also built to run on raised rails and actually wrap around the edges of the rails making derailment almost impossible, and they are considered the safest and most comfortable of any high-speed transit system. The issue with this is that the magnets that supply the levitation and propulsion are built into the rails, making the rails more expensive than the steel rails that normal and high-speed steel wheel trains run on. This results in a system with very high initial costs but extremely low energy usage and operating costs. There is hope that like every other new technology, costs would come down as more systems are constructed.

High-speed travel between American cities is within our grasp, the technology, even for maglev, is established and in use in other parts of the world. But it will require a commitment of federal funds to sponsor the development of the needed high-speed track (steel rails or maglev rails). And that is the key: sponsor, not own. We don't need another Amtrak requiring the federal or even local government to help make ends meet. We need to provide funds to developers to help them build high-speed rail systems. But those developers will then own, operate, and maintain the track as part of a public trust. The trust will receive revenues generated by the trains and clearly publish the maintenance costs and those charged by the mile to any company that chooses to use the rails. Private investment in the rails would correlate to the investment and return on investment that they would receive from the public trust. Trains operated by companies then pay the developer to use the rails. That way, multiple train operators can operate on the rails, which are independent non-profit entities caring for the rails under the supervision of the Federal Railroad Administration. It would be similar to the way that the Federal Aviation Administration runs the skies to let multiple airlines operate through the same airports.

We need to integrate these new rail systems into our existing cities and airports. There was a proposed train from Houston to Dallas, a route of 240 miles, with only three stations. It wouldn't connect with airports, its intermediate stop was halfway between two small cities, and in Houston, the station was going to be located seven miles from downtown. In contrast, the 320-mile-long Tokaido Shinkansen (Japanese high-speed rail) that runs from Tokyo to Osaka has seventeen stations, some in the heart of major cities. The limited stop train makes the run in two and half hours, stopping at six of the stations. The slowest train makes the run in four hours and stops at all stations.

Instead of a major city to major city three-station development, we need to look at integrated rail developments that take advantage of our airports and connect our smaller cities as well. While airplanes take a great deal of effort to tie in small cities between their major destinations, for trains, it is simply stopping, letting people on and off, and then accelerating back up to speed. Linking these intermediate points into the system is the strongest advantage that trains have over airplanes and should be exploited to its fullest potential.

THE TRANSPORTATION POLICY

- Allocate a portion of the infrastructure funds generated by the BTU tax laid out in the chapter The Realistic Green Deal to fund the construction of urban transit systems and highspeed long-distance transportation between metro areas

24.
The American House

The best thing about the future is that it comes one day at a time.

—Abraham Lincoln

Older American homes were built during a time of abundant and cheap energy. Energy efficiency was not a priority, especially in homes built for less wealthy Americans. At some point, we will need to move toward using less fossil fuel. The more energy we need, the more expensive the conversion will be. Reducing energy consumption and becoming more energy efficient has enormous benefits, both now and later. That means it is time for the American house to change as well. The need is for more and better thermal insulation. It means changing from fossil fuels to more efficient electric heating and cooking systems. Studies have shown that making homes more energy efficient increases costs by 5–10%, which is easily recovered by reduced energy bills. And most new homes are much more energy efficient; some have even been built with enough renewable energy to be zero carbon emission homes, but the question is what to do about all the existing homes?

That is going to present a problem to those same less wealthy Americans who don't live in energy-efficient houses. This needs to be addressed; these homes make up the bulk of American homes, and making them more energy efficient not only improves the lives of those living in them but reduces the energy that has to be converted from hydrocarbon to renewable. Converting these existing homes involves multiple options. Adding thicker/better insulation to the wall and attic can cost around $2,000; adding new double-paned insulated windows costs about $120 per window; and compared with old single-pane glass, it is a major improvement. The problem is low-income families may not be able to afford these changes. This is an area where technology solutions are readily available and easy to implement, but paying for it is the issue. And the two major areas to address will be low-income homes owned by residents and low-income rentals (homes and apartments), many of which are government subsidized.

For low-income homeowners, an incentive or tax break may help them in improving their homes. As the homeowner, the money they spend comes back as a benefit with reduced electric and heating bills. The problem is coming up with the cash to fund the energy improvement. There may even be some interest from utility providers to provide cash to help poor homeowners when the utility can receive a credit for the money they spend. This has two benefits: they get a tax break on what they spend to help poor homeowners, and they get reduced power demand, thereby reducing the likelihood of rolling blackouts and the need to build more power plants.

The problem is when it is a rental property. The owner has less incentive to improve the building when the renter is paying the utility bills. Especially when federal subsidies are involved for renters, this makes improvements on or construction of new low-cost rentals unappealing to most investors without federal money. So, what to do about poor apartments?

There are some problems that simply should not be solved by the federal government. Solutions needs to be provided at a local level with local involvement but with federal loans. Remember, with massive federal debt, the government needs to stop passing out federal dollars with zero return on investment. Local governments are better suited to identify their needs and solutions to resolve them, but they need Federal oversight to prevent and limit corruption and waste. There also needs to be consideration of where the low-income housing is located. Most low-income housing is in low-income areas away from where the bulk of the jobs are at. This puts pressure on the low-income workers to have some form of transportation to the areas where the jobs are. Local areas need to look at creating a mix of housing and locating these new energy-efficient apartments in middle- or upper-class areas to move low-income workers closer to where jobs and better schools are located.

This effort has traditionally been fought by middle- and upper-class residents who fear the crime and poverty that is thought to accompany low-income housing. The reason this needs to happen is that in the changing energy economy, we need workers to be closer to their jobs to reduce the energy requirements for them to get to those jobs. There are also benefits in diversifying neighborhoods and schools in those areas. The plan should be to build complexes with a few low-income units (and could have middle-income units as well in the complex) and to scatter these complexes across areas so that low-income apartments are not concentrated. The unfortunate history is that large complexes of predominantly low-income housing have become troublesome areas. The low-income complexes don't generate revenue for the owners, they come with federal subsidies that require a lot of paperwork, and as a result, the owners don't invest enough maintenance or upkeep. Large low-income complexes need to be avoided until we understand why they tend to become troubled areas.

If this seems like a plan to break up areas of poverty, it is. Concentrated areas of poverty have proven extremely difficult to correct with grants and distribution of government funds. What breaks cycles of poverty is better education and jobs, but how do you convince employers and educators to relocate to lower-income areas? The answer is to not only bring the jobs and better schools to those areas but to move the working classes to the places with jobs and better schools and in small groups so as not to disrupt the areas that people are being moved to.

Some urban poor areas have undergone a revitalization as middle classes have moved in, the so-called gentrification that people sometimes fear. The fear is that as property values rise, property taxes go up as well and end up pushing poor people out of an area. The trick is not to have property taxes rise for everyone and create the push. It is common practice to stop increasing property taxes for people over a certain age to keep the elderly from being pushed out of their houses. We need to do the same for poor people and low-income complexes. Property taxes need to be based on specific houses and occupants, rather than broadly increasing the assessed values of houses in a neighborhood. This may seem counterintuitive, but it is not the middle- and upper-class households moving in that causes people to be pushed out, it is the rise in property taxes driven by the improvement of neighbor's houses that provides the push. We need to focus on keeping areas diverse from an income perspective.

This should apply to apartment complexes to encourage owners not to raise rents for low-income families. In the chapter on basic income, we will look at how to help poor families make ends meet, but that does no good if proprietors raise the rent and all that assistance goes into housing. The idea is for apartments for low-income renters to get a lower property tax, but that only applies if they maintain low-income rents. Raise the rents, and property taxes go up.

The reason is that in the changing energy landscape, areas need to become self-sufficient and able to support a diversity of jobs in a region. By pushing the poor out of a neighborhood or city center, low-income jobs can't be filled except from outlying areas. More diverse housing is needed to prevent that. Not only does it help from an energy standpoint, it also helps from a social standpoint in that low-, middle-, and upper-income families all living in the same area, all sending their children to the same schools, and all working and using the businesses in an area breaks down societal barriers that are created when people don't interact with those outside of their income, race, or religion. America must either accept the diversity in the country or suffer the consequences. The government gridlock and societal ills are partially a result of this county rejecting diversity.

This issue also extends into the layout and design of American cities. Too many of our cities and suburbs are laid out with cars as the sole means of transportation. While the present thought is to convert all cars to electric, a better solution is to make walking and cycling the mode of transportation. It gets back to the idea that even an electric car is a 3,000-pound device that makes moving a 180-pound person around very energy inefficient. When you look at older areas in American and European cities, the original layout was focused on cycling, walking, and horse riding. Those same narrow streets that are so unfriendly to cars stemmed from a design that was created before cars were even present.

New areas in a few American cities are focusing on that same style of mixed-use areas with walking and cycling as part of their design. These areas are part of a movement called New Urbanism, an urban design effort to develop more environmentally friendly areas. These are focused on creating walkable neighborhoods with a wide range of housing and businesses that interconnect with mass transit.

The question is what to do with cities and suburbs that were laid out during the age of cars. This goes back to inertia. While we may desire rapid change to a "green energy" future, we are going to have to accept that areas designed based on the "old energy" rules are going to take time to change. We need not outlaw the old rules, but to make those unwilling to change to pay a higher cost for choosing to live according to the old rules. Want to live in an energy-inefficient home in a suburb that doesn't support walking or cycling? You can, but it will be a more expensive lifestyle when fossil fuels become scarce.

THE HOUSING POLICY

- Work with electricity and natural gas providers to set up a program to help low-income homeowners to improve their houses energy efficiency;
- Work with local governments to adjust property tax assessments to reflect the homeowner's and renter's income levels;
- Loan federal money to local governments to support the development of disbursed energy efficient low-income rental housing.

SECTION V: WELFARE POLICIES

Welfare: (noun) the general health, happiness, and safety of a person, an animal, or a group. Synonym: well-being.

—OXFORD LANGUAGES

Never leave that till tomorrow which you can do today.

—BENJAMIN FRANKLIN

25.
Social Security

Social Security is based on a principle. It's based on the principle that you care about other people. You care whether the widow across town, a disabled widow, is going to be able to have food to eat.

—NOAM CHOMSKY

Social Security was created in 1935 during the Great Depression. There had been several attempts to intervene before 1935 by populist politicians and reformers who wanted some form of guaranteed income for the unemployed and the elderly. From the infamous Louisiana politician Huey Long's promise of confiscating the wealth of the nation's rich so the federal government could provide an annual income to poor and middle-income families to Dr. Francis Townsend of California who proposed a 2% national sales tax to fund a pension to everyone over sixty years old who had retired, Social Security wasn't a revolutionary idea. While many will call it a reaction to the Great Depression, the world changing from an agrarian to an industrial society was a major factor. For farmers, the retirement plan is the farm. The land and the extended family form a means of support when they grow old. Even in cities, families that owned businesses and lived in multi-generational houses had their form of retirement plan through the extended family and the family business. For industrial workers renting apartments in cities, there was no extended family or accumulated wealth that could be used as a retirement plan. While a few companies offered pensions in the 1930s, the industrial pension was really a development of the post–World War II period as a way of attracting and retaining workers. In the 1930s, there were very few, and therefore most workers had no retirement income. The crash of 1929 was followed by 22% unemployment, a decreasing money supply, and extremely hard times for the nation's poor, which made that lack of retirement plan even more serious.

The Great Depression pushed President Franklin Roosevelt to form the Committee on Economic Security in 1934 that would perform a comprehensive study and recommend a solution for the unemployed, disabled, and the elderly. The Social Security Act was submitted to Congress in January 1935 and signed into law in August 1935.

The basic concept is that workers pay into Social Security via taxes taken out of their paychecks (payroll taxes paid by employee and employer) and that money goes into the Social Security Trust Fund. When the worker retires, he gets a monthly payment based on his salary during his time working. That is the concept, but the reality is that Cost of Living Adjustments (COLAs) are necessary to increase payments to keep up with increased costs of food, rent, utilities, and life in general. This means that the money contributed will not fully cover all future payments, but that is okay, as long as young workers are also paying into the fund, right? Wrong. In 1977, a set of amendments was passed to address short-term and long-term problems. The short-term problem was that a bad economy had decreased payments into the trust fund, and the long-term problem was that people were living longer and COLA adjustments had eaten into the money in the trust fund. The payroll tax was raised from 5.85% to 6.7% over the five years between 1977 and 1982; benefits were reduced slightly; and the yearly COLA adjustment was increased. These fixes were meant to keep the trust fund solvent for fifty years. They didn't even last ten years. By 1983, more amendments were needed. They raised the retirement age from sixty-five to sixty-seven, not taking full effect until 2027, they made Social Security benefits taxable, increased payroll tax from 6.7% to its current 7.65% (6.2% for Social Security and 1.45% for Medicare) over the course of the next eight years, and numerous other tweaks. Which brings us to the present.

As of December 2021, the Social Security Trust Fund (SSTF) had $2.85 trillion. Now that money is in federal securities and considered part of the "intragovernmental funds" of the federal debt. In other words, the

SSTF has purchased $2.85 trillion of US Treasury bonds not available to the public. Funding-wise, 2021 was a losing year for Social Security. It brought in $1.02 trillion from the Social Security part of payroll taxes (90%), interest on the $2.8 trillion trust fund (7%), and taxing the benefits paid to retirees (3%, remember that change in 1983 to help keep the program solvent). The Social Security program paid $1.14 trillion to benefit payments (98.7%), administrative costs (0.6%), the Railroad Retirement Financial Interchange (0.5%), and had to draw down the trust fund by $56 billion (0.2%). That's the good news. The bad news is that the 2022 Social Security trustees report pointed out that in 1955, eight workers paid into the program for every retired worker. In 2020, that has dropped to 2.8 workers per beneficiary and is expected to further decline till its equal to two or fewer workers per beneficiary by 2037.[61] The issue is that people are living longer, more people are retiring, and the labor force is shrinking. The trustees report predicted that costs would exceed income and interest payments for the foreseeable future, and that by 2031, the trust fund would be depleted to about $1.25 trillion and that without change sometime after 2034, the trust fund could be exhausted. Also remember the trust fund is part of the national debt. As the Social Security Administration (SSA) draws it down, the US Treasury has to replace it by issuing more public debt to replace the drawdown.

With everything else going on, it's going to be nearly impossible to get lawmakers to focus on a problem that they may think is ten to fifteen years away, especially if they think they won't be running for office at that time. But the problem is that by the time you wait until, say, 2030, the trust fund will have been depleted by a third, and the interest income that was making up some of the difference has also decreased by a third. While politicians may want to wait for this to become a crisis to deflect the voters' anger when they either need to raise taxes or cut benefits, it is best to deal with the problem as early as possible.

So, what is to be done? The Heritage Foundation, a conservative think tank, favors cutting the benefits and siphoning money from payroll taxes into individual investment accounts. That would benefit high-income wage earners, as Social Security provides a greater benefit to low-income earners than it does for high earners. The way it presently works is that the SSA takes the average of the best thirty-five years of income and comes up with a monthly salary referred to as the Average Indexed Monthly Earnings (AIME). The AIME is then adjusted based on "bend points." Up to the first bend point, the retiree gets $0.90 for every dollar of AIME, from the first bend point to the second, the retiree gets $0.32 for every dollar of AIME, and above the second bend point, the retiree gets $0.15 for every dollar in the AIME. If that sounds complicated, it is. The table below shows the impact based on average yearly salary for the best thirty-five years of income and the resulting yearly salary the retiree would get for 2020.

35-Year Average Income before Retirement	Monthly income from Bend Points (for 2020)			SSA Yearly Retirement Income
	1st	2nd	Above 2nd	
$10,000.00	$750.00	-	-	$9,000.00
$20,000.00	$833.40	$237.01	-	$12,844.96
$40,000.00	$833.40	$770.35	-	$19,244.96
$60,000.00	$833.40	$1,303.68	-	$25,644.96
$80,000.00	$833.40	$1,786.56	$162.55	$33,390.12
$100,000.00	$833.40	$1,786.56	$412.55	$36,390.12
$150,000.00	$833.40	$1,786.56	$1,037.55	$43,890.12
$200,000.00	$833.40	$1,786.56	$1,662.55	$51,390.12

TABLE 8. SOCIAL SECURITY BENEFITS

Now, these numbers assume that the person retires at full retirement age (sixty-six for those born before 1960, and sixty-seven for those born after 1960). People can opt to retire at sixty-two, but they receive only 72.5% of the benefit. Someone retiring at sixty-seven, who made less than $12,000 per year, would get 90% of their monthly salary every month for a less than $700 annual donation. A person who made an average of $40,000 for at least thirty-five years before his retirement gets almost half, and depending how long he lives, is likely to get more out of Social Security than he put in. For the middle-income earners, it's less of a deal, but remember everyone was only paying in 6.2% and their employer was kicking in another 6.2%. The guy who was making $100,000 before retirement will get 36% of his pre-retirement income for only a 6.2% investment. For the high earners, it's still not a terrible deal, because you need to realize that Social Security taxes cap out. For 2019, that was at $132,900 of income. The person paid $8,239.80 in Social Security tax, and any amount above that was not subject to Social Security taxes. Again, how long the retiree lives determines how good of an investment it was, and the benefits are transferable to spouses and children. Conservatives may want to divert this money to personal investment accounts, but for most low- and middle-income earners, this is the best bet, especially because it requires the employer to match their donation. The Heritage Foundation claims that an average retiree would have twice as much income in retirement if they invested it themselves. This is not entirely true if you only consider the 6.2% direct cost to the employee. The facts are that for employees making $50,000 or less, Social Security is a good deal. It is only for the high earners that it becomes less clear and even then, it depends how long they live. This is because after someone retires, they start eating into their investment's principal cash, meaning that if they live too long, they might be out of cash. Social Security pays per month and has cost-of-living adjustments. This makes it the better deal, but this is also the reason Social Security is in trouble as it is producing better results for retirees, but without enough new young workers paying in, it may not last.

It is the company's 6.2% contribution that people view differently. Some view it as part of the worker's wage that they are owed, and at the moment, it is considered company money that is not given to the worker. Nor does it count against the worker's income taxes, so that 6.2% changes the equation if it is viewed as part of the worker's salary rather than company overhead.

Fun fact: When ride-share companies (Uber and Lyft are examples) and other employers of the "gig economy" label their workers as contractors, it makes the workers into self-employed per SSA, and the workers then have to pay the entire 15.3% of payroll taxes (12.4% is social security) and the companies are off the hook. This makes life simpler for companies and improves their bottom line at the cost of the workers having to pay more tax. Which brings us to the issue of the self-employed who are having to pay the entire 12.4% alone, and if they are making $50,000 per year before retirement, they need to live about eighteen years on Social Security before it makes more sense than personal retirement accounts.

There are a lot of assumptions going on in the calculations above. Typically, everyone starts out at lower salaries that increase for inflation, but as people age, they move into more experienced and better paying jobs. That means that the initial payments into personal retirement accounts are small but have the most time to gain interest. To further complicate matters, what was the assumption on interest rates? Did you account for a couple of recessions? What rate did you use for that person's salary to increase at? Where and when did it plateau? These and other factors create an array of possible outcomes. The general rule is Social Security works best for low and middle incomes who live about fifteen or more years after they retire when compared to personal retirement accounts.

The facts are that at this time Social Security pays out benefits to 55 million Americans, and 42% of older Americans rely on Social Security for at least half of their income. In 2021, more than a trillion dollars passed through the Social Security Administration, and trying to switch that to some kind of personal retirement account while still guaranteeing the payments to the existing 55 million retirees would be near impossible and could be a massive disruption to American society. As a result, Social Security has become untouchable in American politics; politicians dare not touch it, lest they face the wrath of those 55 million elderly voters. But in an era of declining workers paying in and increasing retirees, there is a real crisis in the not-too-distant future, and something needs to be done.

The options are to increase the Social Security tax, increase the salary cap, or decrease the benefits, or change the bend points to reduce the benefits. At present, Democrats in Congress are proposing to raise the salary cap on the Social Security tax for high earners. That would provide some additional funds to the SSA, but it would also make Social Security less of a deal for people making over $132,000 per year. Tax increases won't be popular, and while some may advocate for changing the bend points and only impacting the high earners, they don't make up enough of the retirees to provide the necessary financial relief. No single solution pops out as better or worse, but some effort needs to be made to ensure that Social Security doesn't become insolvent in only fifteen years, because the way the laws are written, any shortfall in Social Security has to be made up from the general funds of the federal budget, and as the next chapter will show, that is already a problem.

The Social Security Trustees have provided reports with a number of possible solutions.[62] The challenge is to get our lawmakers to sit down, actually review, and agree on a course of action that doesn't create a crisis in 2035.

The Social Security Policy

- Immediately convene a bipartisan committee with the mission to determine the best way to keep Social Security solvent into the future.

26.
Medicare and Medicaid

No one was elected to Congress because he or she promised to cut Social Security, Medicare, or Medicaid.

—James P. Hoffa

Before diving into Medicare and Medicaid, we need to understand mandatory and discretionary spending in regard to the federal budget. Mandatory spending is spending required by law, and at this time, it encompasses the three big social programs of Social Security, Medicare, and Medicaid. It also includes interest on the national debt, retirement programs for federal employees, unemployment benefits, and some other programs where a law directs the spending. Discretionary funds are everything left over and require an appropriation act to authorize spending. Defense spending and the various cabinets, are discretionary spending.

Social Security is almost covered by the payroll taxes it takes in and the benefits and administrative costs it pays out. As discussed in the previous chapter, it has a $2.8 trillion trust fund made up of US Treasury notes and is considered part of the national debt. Medicare and Medicaid are a different story.

Medicare began in 1966 as a national health insurance program for people sixty-five and over. It was later expanded to cover younger people with disabilities or some specific diseases. It is funded by payroll taxes and premiums paid by those who enroll in Medicare, and then any shortfall is made up by either the trust fund established as part of the program or by federal appropriations. From October 1, 2013, to September 30, 2019, Medicare costs exceeded payroll taxes and premiums by $2 trillion, the federal portion of Medicaid cost $2.2 trillion, all of which came from the federal budget. During the same time frame, the national debt increased by $3.9 trillion. If the costs for Medicaid were reduced to zero and Medicare was able to actually operate within the payroll taxes and premiums, the United States would have had a surplus of almost $300 billion.

That isn't a fair assessment. The federal portion of Medicaid was always considered part of the federal budget, but it is an example of how skewed things are. Unless Medicare and Medicaid can be brought under control, there is simply no means short of massive tax increases that the federal budget can be balanced. One of the unspoken problems with Medicare is the changing demographics of the American workforce. People are living longer and relying on Medicare for more years. Add this to the reduced young workforce paying Medicare taxes to support the system, and you can see how things have gotten to the point where payroll taxes can't keep up.

There is another problem. Medicare has become increasingly complicated and torturous as can be seen by the number of books, podcasts, and paid advisors willing to help the elderly get "as much out of Medicare as you rightly deserve." An entire industry has sprung out of Medicare billing and negotiations between the Centers for Medicare & Medicaid Services (CMS), which is the federal agency within the United States Department of Health and Human Services (HHS) that administers the Medicare program, the health insurance providers they contract with to run the system and actually deliver the benefits to the participants, and the healthcare providers about the costs that will be accepted.

On a side note, those billing specialists are using the Healthcare Common Procedure Coding System (HCPCS), which was started in 1978 when the *H* stood for Health Care Financing Administration (HCFA), the CMS predecessor. The HCPCS has three levels and comprises about 7,000 codes that cover every procedure, medical equipment, diagnostic test, and even transportation service that Medicare covers. There are college courses just to understand and use it, never mind the books that abound to try and decipher it. Because in 1996, the Health Insurance Portability and Accountability Act (HIPAA) made these HCPCS codes mandatory for all healthcare information and for all payment by Medicare or Medicaid. This does provide some uniformity, no matter which doctor a Medicare patient visits for an allergy injection (HCPCS code 95115), the medical record will reflect the same code and that doctor will be paid the same amount as another doctor in that same geographic region for that same service. Federal bureaucracy at its finest.

How have these cost-cutting efforts worked out? Not well. The data shows that Medicare costs have been rising. Looking at the costs between 2014 and 2019 and averaging gives us a rate of increase of about 5.4% per year. Now, there has been a 2.78% per year increase in the number of Medicare recipients, 2014 had just over 54 million people using Medicare. By 2019, it was 61.5 million. But costs per recipient have been increasing; inflation averaged about 2% per year for the same five years' the costs for Medicare have increased by 2.7% per year. Some of this has been offset by increases in premiums, but even with the additional premiums, the costs have increased by average 2.3% per year. Which totals up to Medicare being on average $370 billion per year over what it takes in for premiums and payroll taxes, which since this is required by law, was taken out of the federal budget ahead of everything else.

A lot of people have argued that the major issue in Medicare is the amount that is spent during the final year of a person's life. Previous studies put the costs at 13.5% to 25% of Medicare, but a 2019 study showed that people in their final year constitute 4–5% of all Medicare participants, but their accrued costs were 20.1% of all Medicare costs.[63] One expert asked, how do you know when someone is in that last year of life and not undergoing a singular event?[64] Which brings up the issue of "death committees" deciding when someone has passed the point of no return. No one wants to have their fate decided by a faceless committee of people they don't know, which has to be taken into consideration.

A 2011 study identified not only the end-of-life group, but two other groups with very high health costs, those experiencing a singular high-cost health event and those with chronic conditions.[65] This study looked at overall spending on healthcare and estimated 18.2 million people (~5%

of population) accounted for almost 60% of healthcare costs overall. The 2011 study estimated end-of-life costs by 2 million of the group at 13.5% of overall healthcare costs, not just Medicare. The singular event group was composed of about 9 million individuals (2.5% of population) who experience a major medical event, but in the following year had very low medical expenses. The long-term chronic group was about 7.3 million individuals and showed a pattern of persistently high medical costs from year to year. We can assume that Medicare has the same pattern as overall healthcare costs—three groups creating the bulk of the costs, the people with singular events, the people with chronic conditions, and those in the last year of their life. Because Medicare serves an older segment of the population, the size and percentages of those groups will be different. While we want a simple solution that reduces costs, it may not be possible because these three main groups have different drivers.

Costs have increased more than inflation. Think about why. Where does the pressure come from to control costs? The program participants don't see their costs; they see their premiums and copays and additional insurances they buy to cover things Medicare won't. There is also a group of experts willing to show them how to get the most money out of Medicare to save their retirement dollars. The health insurance people are just acting as a passthrough between CMS, the healthcare providers, and the participants, all while making a profit. The healthcare providers are in a constant struggle with the fixed costs per billing item, and as a result, they have hired specialists (you can get a college degree as a medical billing specialist) to ensure they bill for every item possible, creating medical bills that participants have a hard time understanding. Which leaves the CMS's effort to control costs when politicians scream over the Medicare price tag, but they are not able to deny services to participants or control what's going on as the laws governing Medicare require them to provide healthcare. Which kicks the whole issue back into Congress's lap to find a solution. And no politician wants 61.5 million Medicare participants mad at him, so Congress authorizes more deficit spending and hopes the whole financial time bomb detonates on someone else's watch.

So, what is the solution when a centralized way to manage healthcare and accompanying costs hasn't worked? The solution is to use a decentralized method, where millions of participants work to keep costs down. The means is to set up individual accounts for each participant, who then have to manage their own health costs. The accounts would be credited with each account owner's share of the premiums and portion of payroll taxes. They would manage their costs throughout the year and any money left over would roll over to next year and get added to their new yearly allotment of money. If they spent all their money, then the program would pay $0.50 per dollar for the first $6,000 spent beyond the account with the participant picking up the rest, any amount over that, the participant pays $0.75 for every dollar and the program picks up $0.25 of every dollar. This would have to be done with all providers adopting a more transparent cost structure and publishing their rates ahead of time (more on that in the next chapter). In 2019, for example, there were 61.5 million participants in Medicare, and taxes and premiums amounted to $413.3 billion. Each account would get credited with $5,000, and $105.8 billion would be held back to pay the government's management costs and pay for those who go over their $5,000. Now, converting everyone at once makes no sense and could become chaotic. Especially those with chronic conditions or nearing their end-of-life costs, who have no hope of maintaining their medical budget and would be exposed to high out-of-pocket costs.

The idea would be to phase this in. All new enrollees would go into the individual account system along with anyone who wanted to switch over. The idea would be that some would recognize that by managing their personal medical costs they can build up their accounts to protect them from singular events or end-of-life issues. This would appeal to those who are healthy and want to manage their healthcare themselves. It is assumed they would be early adopters of this new system.

Money would be able to transfer between accounts. If someone needed a new kidney, his friends could then transfer money to help him. Money left over when a loved one dies? Participants could name a beneficiary, and they will get the money transferred into their health account. If managing the account is a little too stressful, or a participant is not tech-savvy, then they could designate someone as account manager—a spouse, son, daughter, friend, or a licensed professional.

With Social Security, a monthly check is provided, and it is up to the recipient to budget to manage costs, exactly the model Medicare needs. Rather than trying to control costs from Washington, DC, put that effort in the hands of 61.5 million users who would have a stake in not letting medical costs bankrupt them (or the country).

This is actually similar to the Medical Savings Accounts and the Health Savings Accounts, which have the same idea but are usually tied to a high-deductible health insurance plan. In this case, the CMS would cover some of the costs for more expensive events. Some insurance providers offer catastrophic health insurance to cover for high-cost events, but at present, these plans are only available to people under thirty. By changing over Medicare to personal accounts, it is likely that the healthcare industry will adapt and offer catastrophic insurance to everyone.

This system depends on transparency in healthcare. It also has the potential to disrupt the healthcare insurance industry since several healthcare providers have been marketing procedures and medications to Medicare recipients without mentioning the costs, as the recipients weren't paying them, the program was. These lost profits and jobs and the threat of change will bring the lobbyists out in full force to block this effort, but lacking a trillion dollars in increased taxes, it is the only way to bring Medicare under control, and without Medicare under control, balancing the federal budget cannot happen.

The big issue with Medicare is what to do about people with chronic conditions. They would have no incentive to move to the new system since their costs per year already far exceed the allotment they would receive.

But what to do about new enrollees—force them to join? Chronic medical issues are a persistent issue throughout the healthcare industry, which have major long-term problems that will be addressed in the next chapter.

Then the question is what do we do about Medicaid? Medicaid is really a program shared between the states and the federal government but managed and operated by the states. States are not required to participate in the program, but all have since 1982. Medicaid recipients must be US citizens or "qualified non-citizens," which covers lawful permanent residents, people granted asylum or refugee status, and several other designations. Most of the designations for qualified non-citizens are required to have a five-year waiting period from when they become qualified to when they can receive Medicaid. Medicaid recipients may include low-income adults, their children, and people with certain disabilities. Poverty alone does not necessarily qualify someone for Medicaid, though some states have expanded their Medicaid programs to cover all people below a certain income. The costs for Medicaid in 2019 was $603.6 billion overall with the federal budget picking up $388.8 billion (64.5%) of the price tag and the states picking up the remaining $214.8 billion. The $603.6 billion was split into three major areas: long-term care, $129.8 billion; payments to Medicare premiums, $20.6 billion; and payments to healthcare providers, $450 billion. Remember the discussion about how many children are born into poverty? Medicaid pays for 43% of all births in the United States.

When you get right down to it, Medicaid is about charity. It is about providing healthcare to the poor, including the young, middle-aged, and elderly. The problem is that we aren't generating enough taxes to cover this charity with all the other financial commitments, and as a result, we run deficits in the federal budget, and it increases the national debt. If we're going to continue this charity, we either need to identify enough taxes to cover this commitment, or change to a lower commitment to provide a more basic health service, but it still needs a dedicated tax revenue stream to fund it. We will look at what options are available for additional revenue streams to fund Medicaid in the next chapter.

THE MEDICARE POLICY

- Convert the Medicare system into health savings accounts through a phased process.

27.
Healthcare in All Its Forms

America's healthcare system is neither healthy, caring, nor a system.

—Walter Cronkite

Health insurance is one of the areas that economists find very strange. This is because you are paying for something that is *not* based on your likelihood to use it. At present, the only question you can be asked if you apply for insurance on the healthcare.gov exchange is your age, your zip code, and in some states, whether you smoke. Smokers pay up to 50% more than non-smokers. The 50% is limited by the healthcare law. Health insurance plans for one company in Texas were checked, and smokers were found to pay 32% more than non-smokers of the same age. A similar company supplying insurance plans in California never asked if they were smokers. But no one asks about people who are overweight, don't exercise, or eat in an unhealthy way, and people who are overweight and lack exercise are more likely to need medical care than people who exercise and eat healthy, but these factors aren't taken into consideration for insurance premiums. That makes health insurance an area in which an individual's actions don't have the full consequence that people incur.

If you drive dangerously and accrue speeding tickets, have had frequent accidents, or any other indications of dangerous driving, then insurance companies charge you more for your car insurance. The insurance companies are calculating the likelihood of an accident, and they use a range of factors: age, sex, and driving history, plus they deduct for good students and based on the car you drive. The auto insurance companies calculate the risk of you having an accident, how much it would cost to repair your car and pay for any damage you might do, and then they charge a premium accordingly. There are no laws restricting what the auto insurance can charge or requiring them to provide coverage if you have a very bad driving history.

And this was the pattern that health insurance companies used before the Affordable Care Act (ACA) was enacted in 2010. Prior to the ACA, it was common for insurance companies to require doctor's visits to determine how healthy you were to determine what premiums they would charge. They were even likely to turn people away due to pre-existing conditions that indicated the possibility of large medical bills. The ACA has helped people with pre-existing conditions to be able to get insurance, but the counterpoint has been that healthy people have to carry the burden for those with pre-existing conditions. And that is true in most countries.

Most developed and industrialized countries—Japan, Germany, France, and Britain, for instance—have national health coverage based on different systems. The constant among all of them is that everyone is covered and everyone contributes, and none of them have for-profit insurance providers. They either have national insurance programs run by the government (single-payer), non-profit insurance providers, or a combination of the two (single-payer for basic healthcare, private insurance as a supplement). In all cases, the overhead costs—the costs that don't go to providing healthcare to the insured people—are less than 5% of what health insurance companies charge. Some countries have only 2% overhead, which means 2 cents of every dollar goes to managers, employees, marketing, and other non-healthcare costs. Everything else goes to providing healthcare. Compare that to the American for-profit insurance companies, which have 20% overhead costs, that is 20 cents of every dollar, goes to non-healthcare costs and shareholder dividends. The ACA limited the health insurance companies to 20% overhead. Prior to the ACA, they could make as much as they wanted.

America spends more on healthcare than any country in the world. In 2018, the United States spent 16.9% of the entire gross domestic product on healthcare. The second highest was Switzerland at 12.2% of its GDP. Most developed countries spend 9–11% of their GDP. According to a study of

2019 health costs, that comes out to $10,966 per American for healthcare, while Switzerland was $7,732 per person and Canada was $5,418.[66] France has one of the highest-ranked healthcare systems for quality and outcomes in the world, and France spends about $5,376 per person, less than half of what the United States does.

So, spending almost 17% of the entire economy on healthcare, far more than any other country, we must have one of the best systems in the world, right? Wrong. We have the highest infant mortality of developed countries. America averages 6.3 infant deaths per 1,000 live births. France has 3.3, Japan 2.8, and Sweden 2.3.

On basic life expectancy, we rank fiftieth.[67] There is a metric called healthy life expectancy (HALE) that measures how many years you really have before you become disabled and cannot enjoy an active lifestyle. In this regard, the United States rates worse than all the developed countries that we consider our peers. We are sixty-eighth in the world out of the 183 countries listed on the Wikipedia page. Japan, Switzerland, France, Austria, Australia, and Canada all perform better. A shocking number of what we consider third-world countries perform better on this scale than we do. The United States has more asthma, cardiovascular disease, kidney disease, diabetes, pulmonary disease, and liver disease than any of our peers.

The estimate for 2019 was that there were 29.6 million uninsured Americans. This isn't the very poor, as they qualify for Medicaid that is paid for by federal and state governments, it's the people not covered by insurance from their employer and unwilling or unable to buy insurance directly. Of the 273 million Americans under age sixty-five in 2019, 159 million received health insurance from their employer, and 69 million were covered by Medicaid or Children's Health Insurance Program (CHIP which is also federally funded). Another 10 million got insurance through the ACA exchanges or via private insurance. The bulk of those not receiving health insurance indicated that it was the cost of health insurance that was holding them back.

That brings up the question of the cost of medical care. Where insurance companies and medical providers negotiate for rates and discounts, none are offered to those who lack insurance, and those discounts don't always apply to the dreaded out-of-network costs that people have come to fear. This is a major problem with American healthcare, where horror stories abound of ambulance rides or medical procedures that created massive medical bills. Overwhelming medical costs are cited in 67% of personal bankruptcy fillings according to an article published in the *American Journal of Public Health* in 2019.[68] Another source indicated that Medicare pays less than $13,000 for a hip replacement, but some hospitals, almost a sixth of those surveyed, charged more than $90,000 for it. One charged in excess of $130,000.[69]

The medical industry is one of the few industries, if not the only one, that provides services without indicating what they cost ahead of time. There is in fact one industry required by the federal government to provide detailed costs and have them agreed on beforehand: the funeral industry. The Funeral Rule was enacted in 1984 by the Federal Trade Commission to end the abusive practice of misrepresentation. The rule defined that a General Price List must be provided in writing to anyone inquiring about a funeral arrangement. No such rule exists for healthcare. For the medical industry, it's like sitting down to a meal and you only see the cost after you've eaten.

The unfortunate truth is that those with health insurance tend to go to whatever hospital or specialist they are told to and not ask the price beforehand for any planned surgery. And those without insurance tend to not get any preventative care or checkups until there is a crisis requiring a trip to the emergency room.

Stand outside the ER of a major medical hospital, and you will be shocked to see Uber and Lyft drivers dropping off more patients than ambulances. Why? Because the cost of an ambulance is about $2,000 when the ambulance is in-network or run by the local fire department. Out-of-network ambulances could run over $8,000. Many private ambulance services are now owned by private equity companies and are purposely kept out-of-network to allow them to charge more.

Estimates are that more than 20,000 Americans die each year from preventable or treatable medical conditions.

THE ISSUES ARE VERY SIMPLE:

- Too many Americans lack access to medical care.
- Too much money is spent on medical care.

So why wasn't this better addressed by the ACA? Dr. Makary states, "Healthcare stakeholders spent $514 million lobbying Congress in 2016," and that major hospital systems are buying up smaller hospitals and private practices.[70] This growth of hospital systems and other providers allows one system to dominate a market and control costs in that region outside of Medicare where the government regulates costs.

A proposed part of the ACA had been a "public option" with a government-run health insurance agency that would have competed with private insurance. But due to lobbying efforts by health insurance companies, it was removed. The truth is that in some areas, only a few health insurance companies offer policies that create localized monopolies. Several economists have pointed out these monopolies and indicated that the health insurance market needs more competition. Of the twenty-five largest health insurance companies by market share, eighteen are privately held companies, and only seven are publicly traded. Their stock prices have increased by 263% in the last five years versus the Dow Jones average which increased by 191% during the same time. Exclude CVS (number five in the ranking), which is the only stock to show a loss in the group (CVS acquired Aetna in 2018 in the middle of the survey period), and the remaining six show a 294% increase in stock price. When the public option was excluded in 2010, some argued that health insurance companies were not profit-motivated. But the seven publicly traded companies average almost a 4% profit margin. Just to be clear, that is just profit margin, not overhead. Insurance companies are limited to 20% overhead by the ACA, but nowhere is it specified how much of that overhead can be profit.

Several experts have looked at other developed countries to see how the healthcare systems that are ranked as best in the world operate and what we can learn from them. In *The Healing of America*, T. R. Reid took his sore shoulder around the world to see what the different healthcare systems would do for it, the costs associated, and how the systems operated. The key takeaways from this were that the most successful countries have the following points in common:

- Universal coverage of at least basic healthcare
- A strong emphasis on preventative care to improve lives and reduce future costs
- Low overhead costs (<5%)
- Some limitations on what healthcare is available
- Government regulation of healthcare costs

The last point will concern some people, but we already have this in the United States. Medicare sets limits on what it will pay for procedures as a way to control costs. That hasn't stopped people from figuring out ways around it. In *An American Sickness*, Dr. Elisabeth Rosenthal lays out the case that as a result of these cost controls, hospitals (and other medical providers) have devised "upcoding" where minor procedures or events are coded at higher levels to allow increased billing costs. An example she cites is of a doctor who needed a simple blood draw for a test and received an almost $4,000 bill for it. The doctor knew enough to challenge the bill, but most patients wouldn't. Upcoding is how simple things become very expensive and allows hospitals to charge more.

The nature of hospitals has changed. Single patient rooms were exceedingly rare in the past. Now single rooms are the norm and the four-or-more larger wards are gone—more privacy, but at a much higher cost. This reflects a key issue in the US healthcare system—a disconnect between the customer (patient) who doesn't see the costs and as a result doesn't provide direct pressure on the supplier (hospitals and doctors) to lower them. There are multiple examples of people without insurance getting outrageous medical bills and then with help from people in the industry, they argue those bills down. Most Americans rely on insurance companies to get discounts and to argue with the medical providers without the patients having to get involved. Remember that 67% of personal bankruptcies involve outstanding medical bills.

Some may also be concerned about limiting what healthcare is available, the so-called death panels that were brought up during the opposition to the ACA. But the reality is that a lot of American healthcare is unnecessary. Dr. Makary in his book showed where some providers were doing "health fairs" for local churches and other civic groups to look for older Medicare patients and offer them procedures to remove plaque buildup in their legs, a procedure that was unneeded in almost all cases. The free-wheeling, Wild West approach means that people with insurance (Medicare, Medicaid, or private insurance) can get almost anything and pay very little for it at the time of the procedure. But doctors can ask patients to get things done—additional tests and procedures just to rule out highly unlikely events (think avoid malpractice lawsuit) or things that benefit the doctor. The result is the unnecessary care and runaway costs that are burdening the country.

In the United Kingdom, the National Institute for Health and Clinical Excellence determines what is required and what is unnecessary. When T.R Reid took his sore shoulder to the British National Health Service, he was told that because it was just sore, they would recommend exercises, but would not perform a shoulder arthroplasty (shoulder joint replacement) since it wasn't warranted. His American doctor had been ready to perform the procedure, which not only came at a cost of tens of thousands of dollars, but also had all the risks that major surgery entails. In fact, in most of the developed countries he visited, the general consensus was not to replace the shoulder joint.

The United States has amazing advanced medical services available, and this can be seen by the foreigners who come to the country for advanced treatment in cancer and other areas. The problem is we do a horrible job on the basics and preventative care. The United States has more specialists than most other countries and less general practitioners (GPs). The specialties are where the big money is, and that is where most American medical students want to go. The result is that Americans visit their GPs less than people in other countries. In other countries, preventative care early on is stressed to keep long-term diseases in check and actually control future costs. If a disease is caught early, it is treatable at a far lower cost than if it is allowed to become worse where heroic efforts are required to turn back the disease. Preventative care also stresses healthy eating and a bodyweight to keep people more active into their senior years, which also reduces health costs.

The solution is a system that provides basic healthcare to those who cannot afford expensive health insurance. We also need a system that stresses preventative healthcare to both help us achieve healthier citizens and to catch and detect serious diseases before they become life-threatening and very expensive. That was the objective of the public option that Congress removed from the final ACA bill. And that need has not gone away. If anything, the pandemic has pointed out its necessity as people lost employment and couldn't pay the full cost of health insurance without their employers' help and lost their coverage.

Further, there is a problem with medical records. In France, you are issued a *carte vitale*, a card of life. It is not only your health insurance card, but it also has a memory chip built inside with all your medical records. Go to a new doctor, present your card, and the doctor has your complete medical history. Think about it. Who in the United States has their complete medical history? It is scattered across all the doctors and hospitals we have ever visited.

There have been attempts to set up a central set of health records, but with little success. In this day of changing doctors, shifting insurance landscapes, and some people moving from city to city, it is something that needs to be addressed. Because the real question is who owns your medical history—you or the hospitals and doctors? And if it is you, why not carry it around on a card or have a central location so you can access it? Think about it. On vacation somewhere, an accident occurs, and you can't speak for yourself. Who will remember your medical history, what you are allergic to, or what medications you took in the morning? Our medical records are important enough to protect through laws and regulations requiring privacy, but not important enough for us to have emergency access to them?

Don't like the idea of a central database with all your health records? Then support a card with a memory chip that has all your records. Don't like that idea? Fine, opt out. Keep having your medical records scattered among individual providers. But for those of us who want a better record of our healthcare and don't want to be beholden to a single insurance company, we need a solution.

The French system works extremely well, according to all reports. Everyone has a card with a memory chip, and several of them get lost. But people know to send them to the central agency for the vital card and over 80% get returned to their owners. We copy so much from the Europeans; there is no reason not to copy this as well.

The last problem with insurance is what to do about people with chronic conditions. As mentioned in the last chapter, most medical costs fall into three camps: people in the last year of their life (~2 million individuals in the 2011 study group), people with a singular medical crisis (some form of medical emergency), and those with chronic conditions. Over seven million individuals (40% of the 2011 study group) showed persistently high medical costs and were characterized as having chronic conditions and functional limitations. The group tends to be older with almost half the individuals being over sixty-five. A 2016 study by the Milken Institute looked at this same group and identified that the main diseases were hypertension, type 2 diabetes, and osteoarthritis, the majority of these tracing back to the increased level of obesity within American society.[71] And obesity is becoming an increasing problem. Since the 1960s, the percentage of obese adults has increased from 13.4% to 42.4% in 2018 per the CDC.[72] The level of severely obese adults (body mass index greater than forty) has increased from 0.9% in the 1960s to 9.2% in 2018. We will look at a long-term solution in a later chapter, but the issue is what to do in the short term and to realize that while obesity may be linked to the bulk of individuals having long-term chronic conditions, there are many who have other conditions not related to obesity.

Those with chronic conditions usually require things that aren't covered by a basic healthcare system. They wouldn't be able to use a medical savings plan laid out in the earlier chapter on Medicare as they would rapidly exceed their budget. The problem is what to do with them. It is the thorniest of issues since few of these people can afford the care they require, which pushes it into a public burden. But is it a burden? Is it merely a necessary charity and common good that a civilization is set up to provide? But how to do it without bankrupting the government or shifting focus away from education, which has long-term benefits to society?

The best idea is to go back to the basics—personal responsibility and accountability. We don't check the body mass index (BMI) for people applying for health insurance, but maybe we should. The cold hard truth is that people with higher BMIs tend to have higher healthcare costs, but the ACA prevents insurance from being priced accordingly. That is an issue since it makes obesity a problem with a long-term consequence, but no one feels any short-term pain to make them adjust their behavior. No one wants to go back to the old ways of prohibiting those with pre-existing conditions from getting insured, but just as we allow insurance to be more expensive if someone smokes, we need to allow obesity to have an impact as well.

Why hasn't it been done? Obesity has a higher prevalence in people of color, lower-income people, and those without a college education. The same groups who are the least likely to be able to afford higher insurance costs. The problem is that without a direct financial impact to these people in terms of increased insurance premiums, we are not providing an incentive to change. And that lack of incentive has a determinantal effect on children. Childhood obesity has been on the rise as well. The CDC indicated in 2018 that 19.8% of children and adolescents (two to nineteen years old) qualified as obese. And again, it followed the same pattern as adult obesity. If the parents were wealthy and college-educated, the children were far less likely to be obese.

The solution for chronic conditions is two-fold. First, health insurance premiums can be adjusted by a small amount for obesity. We can start with a small amount (5%) and then adjust based on recommendations from insurers and medical professionals in the future. The second is to have people certified as afflicted with a chronic condition and then allow them to enroll in Medicaid if they are lower-income or be considered at a lower income level for the basic healthcare plan.

And no matter what system you set up, how do you protect it from abuse? There are countless examples of people being certified as disabled, receiving benefits, and then being caught on hidden cameras mowing their yards, playing golf, or participating in other physical activities that they had certified they couldn't perform. It's an unfortunate reality that some will try to game the system. The solution is to have two doctors from independent practices to certify a person as chronic. Maybe one is a government-employed doctor (local VA, public hospital, etc.) randomly chosen to do the certification. That way we are less likely to have healthy people certified as chronic.

THE NEEDED SOLUTION STARTS WITH SOME BASIC STEPS.

First is transparency. Healthcare providers need to provide their costs clearly in writing and readily available. Recent articles have documented how some hospitals are hiding their costs from internet searches. Providers also need to clearly show the in-network/out-network costs.[73] Some providers are already getting rid of the in/out cost difference and moving to an easier and simpler pricing system. Knowing what a procedure will cost and how much people have to pay is the first step in managing costs. This is necessary to allow people to manage their own healthcare, especially if we begin to convert Medicare to individual accounts.

Second is competition. Multiple polls have found that most Americans support a public option.[74] We already have a federal bureaucracy in place managing Medicare, Medicaid, and other federal health plans. There is no reason not to leverage that infrastructure and place an option independent of private insurance. Especially as Medicare operates with very little administrative costs, less than 3%, some years less than 2%. The idea is that it would provide basic health insurance that would include regular doctor visits for preventative care, basic medications, and emergency care for those singular life events as discussed in the chapter on Medicare. It would not provide organ replacement, joint replacement, elective surgeries, expensive cancer care, or anything else deemed as not basic. It would include dental and vision since it makes no sense to provide basic medical care without those as well, but it would be basic—no designer frames, no realignment of teeth. Costs would be controlled by Medicare and there would be no in-network/out-of-network. The cost per person would be on a sliding scale based on income and assets and adjusted if they were deemed to have a chronic condition.

This is not Medicare for all as some progressives have called for, rather it is an option that allows people to take advantage of the Medicare rates that their government has negotiated for a basic healthcare option. That won't appeal to everyone, but it provides an option for healthy people without insurance to get basic care at a reasonable price.

Australia offers a similar system where the government-supported public healthcare provides the basics, and lots of people opt to purchase private health insurance (sometimes with their employer's help) to pay for the healthcare that the public option does not provide.

Third is set up a national medical records system, either carried by the patient on a memory chip or maintained by the government, and all providers have access to it with the patient's permission. The government already maintains databases on Medicare, Medicaid, and the Veterans Administration hospital system, integrating across all three that includes other citizens should be doable, might be painful, but there is no reason it cannot be done. People can opt out if they so choose, but it should be the default that everyone gets included in a national health records system tied to a card with a memory chip. Health insurance providers can then include their information on the card and the American "*carte vitale*" could easily take the place of all the insurance cards, Medicare cards, and other cards that litter our wallets.

Fourth is that the additional funding necessary (for this public option to supplement the premiums for low-income people and to fund Medicaid) should come from a variety of taxes on things that cause some of the medical problems in America--tobacco, alcohol, fast food, junk food, and cannabis.

Taxes on tobacco would increase per recommendations from the Congressional Budget Office that estimated an additional $4 billion could be generated.[75] Taxes on alcohol presently only generate $10 billion on sales of over $250 billion. It is time to raise the alcohol tax and use the additional revenue to support the public option and help finance Medicaid.

A tax on fast food and junk food has been kicked around and discussed for years, but now it is time to implement it. With sales of almost a trillion dollars combined in fast food and junk food, we can expect to get a hundred billion dollars from this area alone.

The last area is cannabis. Many states have legalized recreational and medical cannabis and are collecting taxes on it. But cannabis remains illegal at the federal level. Increased smoking of cannabis has health impacts down the road just as tobacco smoking has, so the federal government needs to get in on the act as soon as possible, remove the drug from federal laws, and enact a tax to help pay for public option health insurance and Medicaid.

28. The Grand Shell Game: Prescription Drugs

The enormous pressures on doctors today to prescribe pills, perform procedures, and please patients, all within a disjointed medical bureaucracy and all with an eye on the bottom line, has contributed to the current prescription drug epidemic.

—ANNA LEMBKE

Prescription medicine is another area where reform is necessary. The problems are two-fold: first, drugs costs have escalated, and some pharmaceutical companies play an elaborate shell game to keep prices high and inflate them as well. The other problem is the over-prescription of drugs. Doctors are too willing to write prescriptions for any complaint. While we are painfully aware of the opioid epidemic that has ravaged American society, most of us are unaware of the over-prescribing of drugs in general. It is not uncommon for people to be on five or more drugs at the same time. Each drug comes with side effects and costs that impact the patient. And the more drugs a patient is on, the more interactions and complications are possible.

So, let's look at the history of insulin. Insulin is normally produced in the body in response to the presence of glucose (sugar). When high levels of glucose are detected, more insulin is released into the bloodstream to counteract it. A diabetic person has either lost the ability to make enough insulin (type 1 diabetes), or their body has become less responsive to insulin (type 2 diabetes). This is a massive oversimplification of the disease, but the key is that prior to 1922, diabetes was 100% fatal. If someone's body lost the ability to regulate insulin, they died. But in 1922, doctors in Toronto,

Canada, were able to extract bovine insulin (insulin from cows) and use it to treat diabetes in humans. Insulin has been used as a treatment ever since.

It has also been modified and mixed with other medications. That is research pharmacology at work. The problem is how companies get paid for that research. The annual cost of insulin has been rising steadily since 1991, but only in the United States. In 2018, a Rand study showed that the average price for a vial of insulin was $98.70 in the United States. The next lowest price was in Chile for $21.48.[76] The prices in most modern countries (Japan, Canada, Germany, France, UK) averaged about $11. There were reports of people driving across the Canadian border to get insulin, and no wonder, in 2018 it averaged $12 per vial.

A 2018 research paper reported that three companies controlled 96% of the global insulin market and that their cost to manufacturer was $2 to $6 per vial.[77] So where was this almost $100-per-vial charge coming from? In other countries, pharmaceutical prices are regulated by the government, but not in the United States. The outrage over insulin prices has reached the point where Congress has acted, but only for insulin, moving to cap monthly costs at $35 for those on Medicare and Medicaid. The uproar over insulin has caused Eli Lilly to announce in February 2023 they were cutting insulin prices by 70%. Novo Nordisk, one of the other giants in insulin, followed suit in March 2023. But even at $35 per month out of pocket, Americans are still paying more for insulin than all other countries.

Why was insulin so overpriced? First, very few companies were making it, meaning there was limited competition. Second, it was required. A type 1 diabetic doesn't have a choice—they either use insulin or they die. Several diabetics have died trying to skimp on their insulin to avoid the costs. And the third reason is that from the 1990s to today, this has become the fastest-growing chronic disease in the United States.

The Center for Disease Control has documented that in the last twenty years, the number of adults diagnosed with diabetes has doubled. Their estimate is that 37 million US adults have diabetes, and one in five don't even know it. Diabetes is the seventh-leading cause of death in the United

States. Type 2 diabetes accounts for 90-to-95% of the cases and is primarily caused by obesity. In most cases, type 2 diabetes can be prevented by losing weight, eating healthy, and staying active. We label this as a chronic illness, one of the leading causes of death, and a major reason why so many people lose an active lifestyle, and it is mostly preventable.

Dr. Elisabeth Rosenthal documents discussions with a spokesperson for Sanofi Aventis, which in 2013 had the bestselling brand of insulin, Lantus. Lantus was scheduled to lose US patent protection in 2015. The spokesperson admitted that the strategy was to optimize profit prior to losing the patent, due to the cheaper generic products on the horizon.[78]

Welcome to the pharmaceutical shell game, the way companies try to maximize profits in the US drug market. Remember, most other developed nations have government-controlled pricing, so for these companies, the US market represents the most lucrative in the world. And they have amazing tools at their disposal.

Patent about to expire? Sue someone. Under the Hatch-Waxman Act, an objection to a generic entering the market sets off a mandatory halt in the Food and Drug Administration's (FDA) consideration of a generic competitor for two and a half years.

Want more than two and a half years? Change the drug slightly and get a new patent. Once that is done and the new drug is being manufactured, shut down the old drug and force everyone to use the new version at a higher price before a generic version of the old drug is on the market.

Dr. Rosenthal recounts a lot of examples in which these and other tools have been used by pharmaceutical companies to hold onto their patents and reap profits from the drugs they produce. Low-cost drugs are repeatedly removed from the market, forcing patients to purchase higher-priced equivalents. There have been public cases where the manufacturing rights of certain drugs have been purchased by companies and then repriced at very large increases. Drugs whose patents are running out are reformulated by the smallest amount to prevent low-cost generics from entering the market.

And that is only the first part of the game.

The second part is the pricing game between insurance companies, pharmacy benefits managers (PBMs), and pharmaceutical companies. Increasingly PBMs have been hired to manage the pharmacy benefits of health insurance plans, either at the request of the employer or insurance company. The problem is the PBM sits like a spider in its web between the pharmacies and the customers. And they have developed something called the "spread."[79] The spread is the difference between what the PBM pays the pharmacy for a drug, the copay it charges to the insured person, combined with what it bills to the employer/insurance company. The result is the profit to the PBM for each dose of a drug, i.e., the "spread." And the PBMs are actively working to increase that profit per dose as much as possible. The PBM manages a list of which drugs are allowed to be prescribed, what the co-pay from the patient is, and what they charge the employer. In some cases, the co-pay will be the actual cost of the drug or even higher, meaning the charge to the employer is all profit. Want to check what the pharmacies are actually paying for the drug? Sorry, the PBMs have a non-disclosure agreement in place and the pharmacy can't tell you.

Well, the good news is we have developed some amazing drugs. Drugs that can help people. Look at the COVID-19 vaccine. CDC.gov can show the vaccination status for each death. For December 2022, unvaccinated people were three times more likely to die than the vaccinated. Now, because almost 80% of the country has been vaccinated, more people are dying that are vaccinated than non-vaccinated in terms of the raw numbers. But when you look at the rates per 100,000 people, unvaccinated deaths are 2.99 per 100,000, while vaccinated are 1.01 deaths per 100,000. In January 2022, when the Omicron variant raged across America, the numbers were even worse, with unvaccinated deaths 34.08 per 100,000, and vaccinated at 3.61. And by the way, if you had the updated booster, for December 2022, the rate per 100,000 people was 0.31.

Modern medicine is truly amazing. The average life expectancy in 1900 was 47.3 years. By 1970 it was 70.8 years. For 2021, it was 76.4 years. Now that's the good news. The bad news is other developed countries have far exceeded that as well as some lesser developed countries. The United States now ranks fiftieth in a comparison of life expectancy with other countries in the world, which relates to our healthcare system and the explosion of obesity, diabetes, cardiovascular disease, and a number of other chronic illnesses.

The second problem is the over-prescription of medications. When patients walk into their doctor's office with a problem, they too regularly walk out with a prescription. Not only that, but most expect to walk out with a prescription. We have come to think that we need to be prescribed something to alleviate the problem, and some people demand it.

The CDC tracks the prescription of antibiotics and has indicated that "at least 28% of antibiotics prescribed in the outpatient setting are unnecessary, meaning that no antibiotic was needed at all." The CDC went on to indicate the "total inappropriate antibiotic use, inclusive of unnecessary use and inappropriate selection, dosing and duration, may approach 50% of all outpatient antibiotic use." The concern is that antibiotic-resistant strains have already developed, and unnecessary use could further the development of antibiotic resistance. As antibiotic resistant strains become more common, doctors lose the tools they need to combat them, which is why over prescription of antibiotics is a concern.

But there is another problem in having patients on multiple drugs at once. A study published in 2019 looked at over-prescription of drugs.[80] It found that "for US adults aged sixty-five and over, 41% were taking five or more prescribed medications during the prior thirty days." It also found that for the age group of forty-five- to sixty-four-year-olds, 18.3% were taking five or more prescribed medications during the prior thirty days. That may not seem like much, but the paper also indicated that "polypharmacy (the taking of five or more drugs at the same time) adjusted for these confounds (interactions between drugs) still leads to increased healthcare costs, more adverse drug events, detrimental drug interactions, increased hospital admissions, cognitive impairment, and more falls."

Another paper released in April 2021 looked at 404 patients with cardiovascular disease and found "polypharmacy, hyper-polypharmacy (ten or more drugs), and severe potential DDIs (drug-drug interactions) are very common in older adults with CVD (cardiovascular disease)."[81]

But before we dismiss this as an old person's problem, remember that 18.3% of forty-five-to-sixty-four-year-olds are on more than five medications, and that was based on 2016 data. New studies are showing that polypharmacy is only increasing, and that even children with chronic problems like asthma are likely to be on multiple drugs at the same time.[82]

The more medications someone takes, the more side effects could impact them and the more likely a drug-to-drug interaction can occur. There are now multiple drug-drug interaction checkers on the internet where you can type in the drugs you are on, and the system will flag any possible interactions. But why do individuals have to check this? Remember the French health system with its memory chip card carrying around your medical records? When you go to multiple doctors, they may not be aware of all the drugs you are on. Monitoring for drug-drug interactions has improved, but the lack of central medical records means that doctors not sharing a common medical network may be unaware they are creating a drug-drug interaction.

So, what do we do?

First, we need to recognize that patent protection for drugs exists so the drug researchers can recoup their costs. Drug patents last for twenty years, and it usually takes about ten years to get the drug to market. The big abuse happens when companies try to extend the patent through reformulation or changing the drug and forcing patients to switch. That needs to stop. A law is needed where if a drug company voluntarily stops manufacturing a drug that is being used by patients, then the patent is void, and any company may pick up the formulation and production of that drug to service the existing market that the drug manufacturer is choosing to exit. The drug company with the patent must notify the FDA a year in advance that it will cease manufacturing a drug to provide time for such a switchover to occur. There are too many cases where people went to their pharmacy only to be told that a drug was no longer available, and their only alternative was a more expensive drug. When patients are dependent on a drug that does not have readily available generics, that is a monopoly for the pharmaceutical company. Any attempt to force patients to switch should be illegal.

Second, when a drug is registered with the FDA, the company also needs to register their cost to manufacturer, including the required overhead, how much they tack on to recoup research costs, and the number of doses required for them to recoup their research costs. The FDA shares that with Medicare, Medicaid, and the VA so they can set acceptable prices for the drugs. Want to raise your price on a drug? Resubmit to the FDA for approval. Does this sound very un-capitalistic? Remember this is most likely a monopoly and the drug manufacturer has a captive audience who may have to buy the drug or die. Charging $100 a vial for something that costs $4 to make is the mark of a monopoly, as was the case with insulin. This isn't like any other market where if you don't like the pricing from one manufacturer, you can go to another. This is the only source of the drug. Now, we need to be reasonable. Initial submittals will be based on estimated cost to manufacturer and to get approval, the drug manufacturer should have the ability to resubmit and not be denied approval once actual costs have been determined, especially when differences are small. But the FDA should also have the means to conduct a surprise audit to ensure that the drug company is accurately reporting costs.

This provides the government the opportunity to step in. Let's say there is a drug that cures a childhood disease, but it is prohibitively expensive. If the drug manufacturer opens their books and shows they are not overcharging, but the drug has a massive benefit, then the government can look at supplementing the costs. It provides the transparency with the drug manufacturers that intelligent decisions can now be made.

The third item goes back to the previous chapter and the recommendation that a central healthcare records system be established. The idea that individual patients need to check for drug interactions in unacceptable. Most people are ill-equipped to perform that level of monitoring. A central health records system or even a portable card would have the ability that every time a drug is added, the system could check, and flag known interactions.

29.
Obesity and the American Farm

We struggle with eating healthy, obesity, and access to good nutrition for everyone. But we have a great opportunity to get on the right side of this battle by beginning to think differently about the way that we eat and the way that we approach food.

— Marcus Samuelsson

American farms use large amounts of fertilizer, which is mainly produced from natural gas. They also use gasoline and diesel-powered machinery. As such, American farms are looking at a crisis when the effort to reduce fossil fuels kicks in. Farms could be given waivers or exemptions, but that only complicates the situation, which is never a good thing. Biofuels should provide some relief on the powered machinery front, but not with nitrogen-based fertilizers. The rise of organic farming and the changing views toward corporate and monoculture farming may offer a solution but are already complicating things.

Monoculture farming is growing a single crop in a large area. The trend in American agriculture is large farms growing massive amounts of a few crops. An example is Iowa, where the bulk of American corn is grown. The view of field after field for as far as the eye can see growing a single crop is actually a new thing. Farms prior to the 1940s grew diverse crops both to feed the farmers and their animals and to not rely on a single crop for the financial success of the farm. Prior to tractors, 40% of a farm's crops generally went to feed the farmer, his family, and his livestock, horses, oxen, and mules to pull the ploughs. No one is suggesting a return to that, but it needs to be recognized that the decline of fossil fuels is going to affect agriculture.

Prior to nitrogen fertilizers, crop rotation was required since most crops reduce nitrogen and other nutrients in the soil. It was necessary to occasionally grow crops that put nitrogen and nutrients back into the soil as well as spread manure from the farm animals on the soil to keep it producing. The Haber-Bosch process discovered how to convert natural gas into ammonia to then produce nitrogen fertilizers. This has allowed the world population to soar. Some scientific papers estimate that up to 4 billion of the world's 8 billion depend on crops grown with synthetic nitrogen fertilizer.[83] At present, the generation of nitrogen fertilizers uses a small percentage of total natural gas consumption, less than 1.5% in the United States and less than 5% worldwide. China is the largest producer and consumer of nitrogen fertilizers. And there is a real concern that the world may not be able to sustain population levels without synthetic nitrogen fertilizer.

One of the issues in the United States is the system of farm subsidies, where the federal government spends $20 billion in subsidies to the largest producers of corn, soybeans, wheat, and rice. These are received by 39% of America's 2.1 million farms. This has two effects. First, it de-risks American agriculture and protects farmers from fluctuations in the price of crops and risk of crop failures. Secondly, it creates an abundance of commodity crops, corn, wheat, and soybeans. The largest of the commodity crops is number-two corn, which can only be used for animal feed, to make the bio-fuel ethanol, or as feedstock for the food processing industry. The food processing industry then takes the feedstock and processes it into individually wrapped food items you find on shelves in every small market or gas station. This is the processed food that the health industry in now pointing to as a major cause of America's obesity.

A third of America's corn crop is fed to animals, a major part of which is beef cattle on giant feedlots. The issue is that cattle haven't evolved to eat corn, they are ruminants that eat grass. Most modern beef cows spend their first six months of life on a pasture with their mothers. After that, they are sent to a feedlot where they are fed that number-two corn, which is inedible for humans. Corn is high in protein and phosphorus and low in calcium when compared to grasses, which is why it is a problem for cows. It takes dietary supplements and medical care to allow a cow to eat corn and not only survive but thrive. So why do it? Cattle fed on corn gain weight faster, and their meat contains more marbling, or fat, along with a more consistent taste. Meat from grass-fed cattle contains more nutrients, but far less fat. There is also the issue of cost. Number-two corn, due to federal farm subsidies, is cheaper than hay and allows for the concentrated feedlots where it is trucked in from those farms in Iowa. The result is that an informal survey on January 26, 2021, found that grocery store ground-beef prices reflect the difference in costs.

- Corn-fed ground beef was $2.33 per pound and had 27% fat.
- Grass-fed ground beef was $5.99 per pound and 15% fat.
- Finally, organic grass-fed beef was $6.99 per pound and 10% fat.

Not exactly a scientific study, but it does show the impact that number-two corn and cattle feedlots have on the price of ground beef. Care to guess what kind of ground beef all the fast-food restaurants are using?

But there are problems with corn-fed beef. Corn-fed beef is higher in fat and doesn't contain the same nutrients as grass-fed beef. And there are concerns that the antibiotics and dietary supplements needed by the cows to eat the number-two corn could leave residues and by-products in the meat. The corn diet causes serious problems for the cows. A lot of them have liver problems and stomach ulcers. When the cows are slaughtered, a lot of the livers are diseased and unusable, which isn't a problem since the livers are discarded. But as cows can grow to slaughter weight in fourteen months, the fact that their liver is failing doesn't stop the process. Beef cattle are packaged meat long before liver problems would have killed them. In contrast, a grass-fed cow takes four to five years to gain enough weight to be ready for slaughter. American farmers sell that number-two corn to the feedlots for about $0.05 to $0.08 per pound when it takes $0.08 to $0.10 per pound to grow it. The difference between those prices are the federal subsidies mentioned earlier. Cheap corn for food, shorter growing cycles (fourteen months versus five years), and able to pack thousands of beef cattle into a small feedlot, those items are the reasons that corn-fed beef sells for almost $4 per pound cheaper than grass-fed beef.

The original farm subsidies started during the Great Depression of the 1930s after the Dust Bowl had wiped out farmers in Oklahoma, Texas, and Nebraska. The purpose of the subsidies was to encourage farmers to not plant all their cropland and leave some land idle or with crops that actually rejuvenated the soil. This was before nitrogen-based fertilizers that could increase soil productivity. It was necessary for cropland to have time to recover from extensive use, and the overuse of farmland was one of the causes of the Dust Bowl. Rejuvenating the soil was done with crops that bound nitrogen into the soil like beans and grasses, which were planted to help the soil recover. Once nitrogen-based fertilizers arrived on the market, that recovery time was no longer needed and was seen as unproductive time when high-value crops could be grown. Those nitrogen-based fertilizers depend on natural gas to be generated, something that needs to be addressed as we move to a less fossil-fuel-based economy.

In the 1970s, farm subsidies changed from encouraging farmers to not over-produce to actually paying farmers to grow more and form bigger farms as a result. This industrialization of the American farm can be traced directly to the actions of the Nixon-era agriculture secretary Earl Butz, who exhorted farmers to plant from "fencerow to fencerow" and to "get big or get out." The reason the Nixon administration did this was that in the 1970s, Russia was experiencing crop failures, and the United States wanted to sell lots of American grain to the Russians and other foreign countries. What started out as food diplomacy enabled the rise of monoculture farms of cheap number-two corn and massive feedlots of cattle, pigs, and other animals, the industrial farming that we now know.

This change in farm subsidies encouraged the giant monoculture farms, but those farms depend on two other major factors. The addition of nitrogen-based fertilizers that keeps the soil producing year after year of planting the same crop. Also, the application of pesticides and herbicides to kill insects and weeds. Some farmers refer to monoculture farming as "spray and pray." Spray the insecticides and weed-killers, and pray that a resistant variant doesn't show up and wipe out the crop. To that end, industrial agriculture companies now offer specialized hybrid seeds that allow stronger pesticides and are more resistant to some pests. These seeds are infertile, meaning the farmer can't plant the corn he grows and expect it to sprout the next year, he has to go back to the agricultural company and buy more. These seeds are touted as producing the largest yields per acre, but they also keep the farmer locked in as a customer for the following year's planting. Monoculture farming and the American industrial agriculture complex are now locked together, which means any change to the farm subsidies threatens some very large companies. Cargill had revenue of $115 billion in 2018, while Archer Daniel Midlands had $64 billion in revenue.

Monoculture farming produces fertilizer runoffs, and there are serious concerns with the pesticides and herbicides being used. Can we sustain this type of food economy in the future, and what are the implications of changing it? The things to consider are the two areas of major impact—cheap beef and junk food.

Feedlots didn't start until the 1950s when Midwestern farmers found they could feed grain, corn, and other foods to animals in fenced enclosures and not require acres of pasture that could be used for other purposes. It took off in the 1970s with the change in food subsidies and availability of cheap grain, mainly number-two corn. Animals in these enclosures eat massive amounts of grain and end up standing in manure and urine. The manure, which used to be prized as fertilizer prior to feedlots, now becomes hazardous waste when concentrated in feedlots. The largest amount of beef cattle was in the 1970s with almost 100 million beef cattle. Today, with the rise of feedlots, we produce more beef with only 31 million beef cattle, less than a third of the 1970s. By getting them to slaughter weight in only fourteen months, you need less cattle to sustain American beef production. This is the source of America's cheap beef at almost $4 per pound cheaper than grass-fed. This problem is not just with cattle; all the animals we depend on for protein have been transitioned to feedlots. And the meat is less healthy, fattier with fewer nutrients, and being so cheap, Americans eat a lot of it. Americans eat almost 140 pounds per person per year of meat which has remained constant in the last fifty years. The mix has changed. In the 1970s, cattle was king, but now chicken has become the most popular meat. But the issue is the same—massive feedlots with animals packed in tightly, an unhealthy situation for the animals that is producing an unhealthy situation in Americans.

Most would consider the grass-fed option, but a German study in 2020 concluded that grass-fed is as bad as or worse than grain-fed from methane emissions and water consumption.[84] This study was limited since it only studied German agriculture and assumed that grazing was not the only source of feed for grass-fed animals and assumed a monoculture type approach as well. The German paper built on past research and even identified that it had not considered carbon sequestration which could be generated through organic pasture grazing. Several studies have considered this and showed that this carbon sequestration could go a long way towards reducing the impact of organic cattle.[85]

But there is a different approach called polyculture where a farmer uses his pasture in a rotational approach and grows cows, chickens, pigs, and crops by rotating them through small pasture areas. This produces high yields, restores soil, and sequesters carbon within the soil. Several farmers using this approach report high yields with little to no fertilizer or antibiotics. The process can also extend into agriculture where planting multiple diverse crops together can reduce the need for pesticides and fertilizers. There appear to be some question as to the impact on the climate, but there is no question that grass-fed meat is less fatty and more nutritious than grain-fed, but the problem is the cost.

When corn-fed beef is combined with processed food there is plenty of data to show that these two factors, cheap corn-fed meat and ultra-processed foods, have contributed to America's health problems. Processed food has become a huge industry and a major piece of it is called ultra-processed food. Ultra-processed food are the foods that have salt, sugar, artificial colors, flavorings, and preservatives added to make them more appealing or make them last longer. One study found that ultra-processed foods make up about 60% of the calories in the typical American diet. Studies have tied ultra-processed food to obesity, diabetes, and several other health issues. Not only is this food unhealthy, but because it tends to be inexpensive, it accounts for a majority of the diet among poor and low-income Americans. The food is inexpensive because it comes from that number-two corn that is federally subsidized. The number-two corn is broken down into its molecular building blocks—carbohydrates, proteins, etc.—and recombined into ultra-processed foods. Ultra-processed foods dominate the shelves in most grocery stores, but especially in poor and low-income areas. While many complain about "food deserts" where healthy food cannot be found, the problem is the lack of economic incentive for food suppliers to bring healthy foods into these areas if the customers are going to continue to buy the cheap, ultra-processed food. Some frame these "food deserts" as a racial inequality, but that looks past the question of economics. Are poor and low-income people going to purchase more expensive "healthy" food rather than cheap "junk" food and fast food? For a Whole Foods, or any other grocery store that emphasizes healthy options, establishing in a poor/low-income area has to be profitable or at least break even. That is going to be a problem if most of the people in those neighborhoods continue to buy ultra-processed foods from the Quickie-Mart and not the healthy grocer. Please remember that there were once grocery stores in these neighborhoods, which have closed or moved due to the purchasing pattern in the neighborhood. And how do you change that?

The problem with sustaining the present food economy is that it depends on nitrogen fertilizers, and it depends on the monoculture farms that are heavily mechanized. It depends on chemicals and antibiotics to raise the corn and treat the corn-fed cattle, and mostly, it depends on federal subsidies. And ultimately, it is producing poor health outcomes for many Americans. We must consider that our present system developed over the last fifty years and isn't likely to be changed with a few laws and a president's stroke of a pen. That food system represents America's preference for ultra-processed foods and fast food from cheap restaurants. The movement toward more organic food has caused some to move away from corn-fed beef and ultra-processed foods, but there is still a huge market for those foods. The grocery store where the informal survey was done had a large selection of grass-fed beef and organic foods, but it had a larger selection of unhealthy food. The problem is that the healthy items cost more, and that may not be affordable for people on fixed or low incomes.

You can't have a healthy civilization without healthy soil. You can't have junk food and have healthy people.

—JOEL SALATIN

An easy answer could be to phase out the federal farm subsidies that produce cheap number-two corn. As the price of corn rises, the corn-fed beef and ultra-processed food will rise as well, creating an incentive for people to eat less unhealthy food and eat more healthy food. But the problem is that it is unlikely to make the unhealthy food cost more or even come close to the cost of healthy food. Even with the taxes on ultra-processed food described in the chapter on healthcare, healthy food will likely remain more expensive. And being more expensive. it will increase the cost of living for poor and low-income families. The other problem is what does it do to the American farmers who were receiving the subsidies? Some family farms are on the edge of bankruptcy, and the lack of subsidies could easily push them over the brink. And while large farms are getting more subsidies, lobbyists for the agricultural industry will not hesitate to make examples out of any family farms that happens to go bankrupt from the loss of federal subsidies.

So as the subsidies are phased out, what do you do? Do you phase in a subsidy to poor families for their food budget, or do you phase in an incentive for farmers to move to an organic or polyculture farming system and thereby try and reduce the cost of healthy foods?

The best way to resolve this is to understand what the goals are. The first goal is to get America away from unhealthy eating habits that are causing health issues. We not only spend more on healthcare, but we also get worse outcomes than all the developed countries in the world, and some of that has to be tied back to the ultra-processed food that makes up too much of the American diet. And while we should not outlaw ultra-processed food, we need to stop any federal subsidies and tax it to provide an incentive for reducing the consumption of ultra-processed foods and feedlot-generated meats. People can continue to eat unhealthily, but we need to make that food cost more to reflect the actual health costs that these foods represent.

The second goal is to move to a more environmentally friendly system of agriculture. Feedlots with their overflowing manure pits, and monoculture farming with its dependence on hybrid seeds, nitrogen fertilizers, herbicides, and pesticides aren't long-term answers. We need to encourage the further development and spread of polyculture farming that relies less on chemicals and fertilizers and is more in step with natural systems in the world.

The third goal is to not increase food costs so that families can't make ends meet, especially poor and low-income families.

The fourth goal is to do this without federal subsidies. Remember we have a massive federal debt to be addressed. Even though we are only talking about $20 billion; it's $20 billion that could be desperately used elsewhere or to pay off the debt.

And finally, we need to consider the American farmer, a bedrock of the conservative movement and a solid backbone to the country. It is going to be a challenge to get America's farmers to change the way they have farmed for almost fifty years, and some pushback must be expected. But several farmers have changed from monoculture to polyculture. Joel Salatin of Polyface Farms has been a champion of the polyculture movements and has published books, YouTube videos, and TED talks to advocate for his system in which cows, pigs, chickens, and rabbits all work together to produce a better farm. One of the lessons from Salatin is that polyculture farming is more labor-intensive and requires more planning and thought. Animals are moved regularly to fresh pasture, sometimes daily. Different animals' natural tendencies are used together to increase farm productivity. We cannot expect American agriculture to change rapidly, but we must put the incentives in place to begin the change. The best solution is to work with farmers on how to change from a federal subsidy for large commodity crops into a subsidy that encourages more ecologically friendly polyculture farming.

The Farming Policy

- Phase out the existing federal subsidy toward commodity crops and phase in a subsidy that is geared towards encouraging polyculture farming.
- Require that the Department of Agriculture work with farmers and experts in polyculture farming to develop incentives that will best do this without increasing the federal budget for agriculture.

SECTION VI: ECONOMIC POLICIES

Economy (noun): 1. the wealth and resources of a country or region, especially in terms of the production and consumption of goods and services 2. careful management of available resources

—OXFORD LANGUAGES

Economic policy is like business—it's all about compromise.

—JOHN DELANEY

30.
The Dreaded T Word

I like to pay taxes. With them, I buy civilization.

—Oliver Wendell Holmes Jr.

Taxes is a word that no politician wants to talk about raising, but faced with a massive national debt, it is a subject that must be discussed—especially at a time with rising interest rates that will cause payments on the national debt to increase. We may not have any choice but to raise taxes as debt interest payments cut deeper into the federal budget. But first we need to understand how tax rates have changed in the last hundred years.

When analyzing tax brackets from the 1920s until now, you must account for inflation. *Inflation* means that things cost more today than they did back then. An example is a gallon of milk. In 1920, milk cost $0.35 per gallon, as the Great Depression started in 1930, milk cost $0.26 per gallon. In 2021, the average price was $3.54 per gallon. By looking at the change in cost for a broad set of things that we use in normal life, economists have calculated the Consumer Price Index (CPI) to determine inflation. You can use the difference in CPI between two times to correlate the federal income tax rates from the 1920s to tax rates for today. In 1933, anyone making between $4,000 and $8,000 was in the federal income tax bracket of 8% tax. Using inflation calculations, we can determine that $4,000/year in 1933 equates to someone making $81,120/year in 2021 and to infer that someone making $4,000 per year in 1933 has close to the same standard of living as someone making $81,120 per year now. The problem is that the person making $81,120/year today now pays 22% federal income tax as well as 7.65% in Social Security and Medicare.[86]

You also need to look at the national debt and how it compares to the Gross Domestic Product (GDP). The GDP measures the economic output of a country and its economic health. It has become one of the critical numbers tracked for national economies by economists and politicians. In 1933, eight years before the start of World War II and during the Great Depression, the national debt was $23 billion, which was 40% of the national GDP. $23 billion in 1933 seems like a small amount, but again, it equates to 40% of the entire country's economic output at the time. As we discuss taxes, national debt, and history, we need to do it in terms of a percentage of the GDP to keep it in perspective.

We start the tax history lesson in 1920 with World War I having recently ended. The federal income tax was introduced in 1913 with the passage of the Sixteenth Amendment. It had been introduced in 1861 to pay for the Civil War but had been repealed in 1872. It was reintroduced in 1894, but a Supreme Court ruling in 1895 called it into question, so the Sixteenth Amendment was required to legalize it within the framework of the Constitution. After having fought World War I and dealt with the 1918 Flu pandemic, the national debt was $25 billion, which was 48% of the GDP. Taxes were high. Someone making $40,000 per year would be in the 27% income tax bracket, and that $40,000 in 1920 translates to an income of about $530,000 in 2020 dollars, which is a 2020 tax bracket of 37%. There were fifty-six tax brackets in 1920, starting at 4% for those making less than $4,000 per year ($54,200 in 2021 dollars) all the way up to 73% for those making over $1,000,000 per year ($13.5 million in 2021 dollars). The national debt was paid down to $16 billion in 1929, 16% of the GDP, just before the stock market crashed and the Great Depression began. As the debt had been paid down, income tax rates had also been lowered. In 1929, the top bracket was $100,000 per year ($1.6 million in 2021 dollars) and was 25%.

Once upon a time my political opponents honored me as possessing the fabulous intellectual and economic power by which I created a worldwide depression all by myself.

—HERBERT HOOVER

The Great Depression saw a massive collapse in the American economy at the same time as a worldwide economic depression occurred. Unemployment skyrocketed from virtually zero to 25.6% in May 1933 and worldwide unemployment rose to 24.9% that year. The global Gross Domestic Product decreased by 26.7%, while the US GDP fell by more than 30%. The 1930s saw an increase in the federal debt and increased tax rates. In 1931, tax rates were increased by President Hoover, a Republican, to 56% for those same people making $100,000 and seven higher tax brackets were added all the way up to those making over a million dollars per year in 1931 with a tax rate of 63%. Tax rates were changed five times during the 1930s as the government tried to address the Great Depression. President Hoover felt that government intervention was not the solution and paid the price for those views as Franklin D. Roosevelt was elected president in 1932. Roosevelt's New Deal federal programs began to put people back to work and ease the financial depression.

With the Great Depression ending and World War II on the horizon, President Roosevelt in 1940 raised military spending in the United States and increased taxes in preparation for going to war. A war that caused the National Debt to balloon from 40% of GDP in 1940 to 118% by 1946. In dollars, it went from $23 billion to $269 billion which was the cost for America to wage World War II. Taxes in America were particularly high during and after World War II to pay down the national debt that had accrued during the war. In 1946, there were twenty-four tax brackets, and the highest was 91% for everyone making more than $200,000. In 1946, $200,000 was equivalent to over $2.5 million today. That person who was making the equivalent of $2.5 million would have been taxed at 56% in 1932, 91% in 1946, and today would be taxed at 37%.

The national debt now is over $31 trillion, over 100% of the GDP, over a hundred times the dollar value of what it was at the end of World War II. And the highest tax bracket in 2023 is 37% for anyone making more than $518,400 per year. Let's understand this: the only reason we are not in a financial crisis over the debt is because interest rates have been remarkably low for the last fifteen years. In December 2008 with a financial crisis in full force and the national debt hitting $10 trillion, the Federal Reserve dropped the interest rate to 0.25%. In May 2023, with the debt at over $31 trillion, the Federal Reserve, in an effort to keep inflation in check, has raised interest rates to 5%, when a year earlier in May 2022, it was 1%. This is going to increase the amount that the government must spend on interest payments on the debt. Because of that and the natural imbalance in the national budget due to social and defense spending, there is not going to be a balanced budget this year or the next. The national debt will not decrease anytime soon unless we raise taxes or make serious spending cuts.

Paying off the national debt has been very difficult. During President Truman's term in office (1945 to 1952) the national debt only decreased by $10 billion, from $269 billion to $259 billion. But as a percentage of the country's GDP, it decreased from 118% to 71%. This was due to the American economy growing in the postwar years. But during this same time, taxes were still very high to help pay down the debt. During Eisenhower's term (1953 to 1960) taxes remained high, and the debt declined as part of the GDP, from 71% to 53% due to the American economy continuing to grow, while it actually increased in dollars from $259 billion to $286 billion. The pattern of the actual debt increasing but continuing to reduce as part of the GDP happened during the Kennedy/Johnson ($62 billion increase in debt), Nixon/Ford ($272 billion increase in debt), and Carter administrations ($287 billion increase in debt) with slight changes to the tax brackets during this period. At the end of the Carter administration in 1980, national debt had declined to 32% of GDP, but had climbed to $908 billion.

The Reagan tax cuts brought sweeping change. The wealthiest people in the country were in a 70% tax bracket if they made more than $108,300 per year ($348,132 in 2020 dollars) when Reagan took office. When he left, there were only two tax brackets. Anyone making less than $18,550 per year paid 15% income tax, and anyone making more paid 28% income tax. The deficit ballooned from $908 billion in 1980 just before Reagan took office to $2.6 trillion and 50% of the GDP in 1988 when Reagan left office.

Reagan almost tripled the debt while he was in office during very good economic times, as the GDP grew from $2.8 trillion to $5.4 trillion. The first President Bush in 1990 and 1992 had to increase taxes as the Reagan cuts had been too deep and were creating too much of a deficit, even though he had promised during his election, "No new taxes." When President Bush left office at the end of 1992, the deficit was $4 trillion and 62% of the GDP.

The Bush tax increases in 1992 combined with a good economy resulted in the Clinton administration having a budget surplus in 1998. It was the last time the United States had a budget surplus. At the end of the Clinton administration in 2000, the national debt was 55% of the GDP, but the amount of debt had risen to $5.6 trillion, again increasing. The GDP had reduced the percentage even though the actual debt was increasing. But this was short lived, the Bush tax cuts in 2001 and the wars in Afghanistan and Iraq restarted the yearly deficits and the increases in the debt. The 2008 Economic Crisis and the associated bailouts and then the 2020 COVID stimulus packages have driven the debt to unprecedented levels.

When we ran up large debts compared to the country's GDP during World War I and World War II, taxes were raised to pay it off. There has been no such effort in the last fifty years as the debt has increased from 31% of the GDP to over 100% of the GDP. In fact, President Trump with a Republican Congress permanently lowered corporate taxes and temporarily lowered income taxes in 2017, and those temporary personal income tax decreases will expire in 2025, with income taxes returning to 2017 levels.

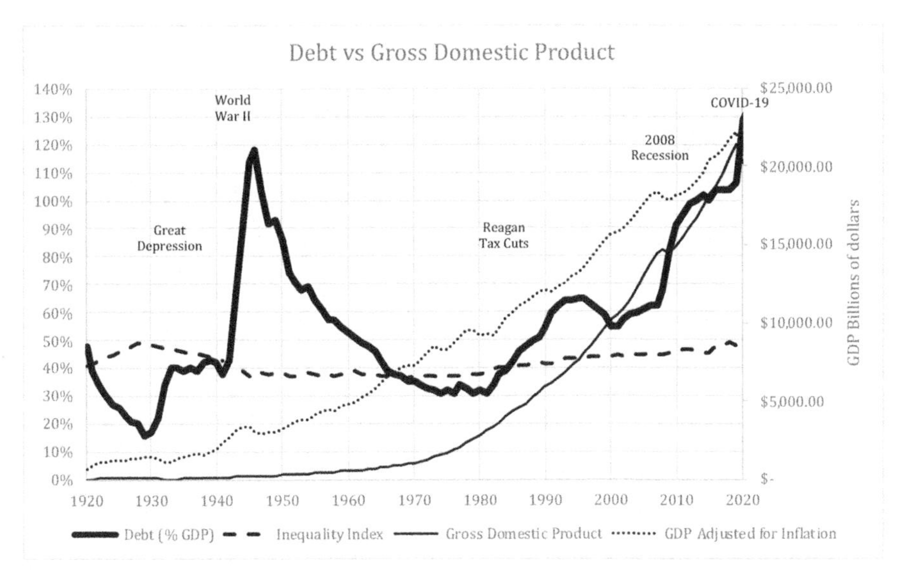

FIGURE 2. US DEBT AND GDP

The chart below shows how tax brackets have changed over various incomes between 1920 and now. This was done by taking all the tax brackets since 1920 and then adjusting the income thresholds by inflation to set up a standardized table. The rates are both income tax and payroll taxes.

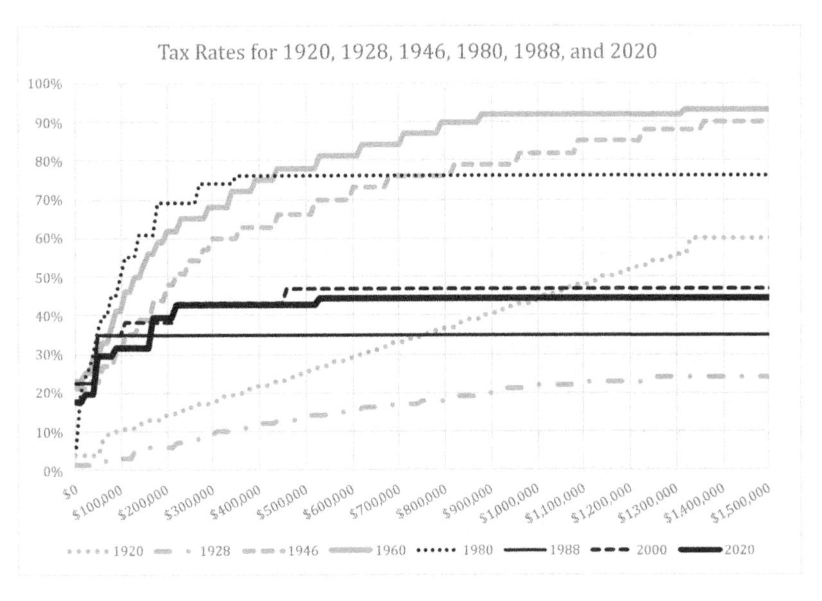

FIGURE 3. TAX RATES OVER TIME

Tax rates in 1920 were fairly progressive, and when the debt from World War I was paid off, they were lowered as the line for 1928 shows. But the Great Depression in 1929 and World War II drove the national debt to over 100% of the GDP, and tax rates were raised to the 1946 line to pay off the debt. By 1960, inflation had pushed more Americans into higher tax rates as the tax brackets were not adjusted for inflation and payroll taxes had increased from 1% during the 1940s to 3% at the end of the Eisenhower administration. By 1980, inflation continued to impact more Americans, and payroll taxes had risen to 6.13% but the highest tax brackets had been reduced to 70%. The Reagan administration took over in 1981, they reduced taxes such that by 1988 there were only two tax brackets.

President Bush raised taxes twice in 1990 and 1992, which resulted in his being replaced by President Clinton who didn't touch taxes, resulting in the 2000 line above. The 2020 line is the result of the President George W. Bush's tax cuts in 2001 and the Trump tax cuts in 2017, putting us where we are today, with a national debt over 100% of the GDP and people making less than $200,000 paying almost (in some cases more than) what they would have paid in 1946 and the wealthy paying a lot less.

What this shows is that our tax structure is no longer as progressive as it was, and this is showing up as an increasing financial inequality. The Gini coefficient was developed by an Italian statistician Corrado Gini in 1921, and it represents how unequal the wealth is in a nation. Zero means wealth is distributed evenly across the populace, while a one means the wealth is concentrated in one person. In 1928 and 1929, just before the economic crash that started the Great Depression, the index was at 0.49. Through the 1950s and 1960s, it declined, reaching its lowest point in 1968 at 0.35. Since then, it has been on the rise reaching 0.5 in 2023. This cannot be totally blamed on the tax structure, but the tax structure is a confirmed factor. Countries with a higher Gini coefficient are South Africa (0.63), Namibia (0.59), India (0.57), Saudi Arabia (0.55), Brazil (0.54), and Mexico (0.49), along with several small third-world countries. The highest value in Europe is 0.35.

The adage is that the rich are getting richer, and this is borne out by the data. Since 1989, the net wealth controlled by the top 1% of American incomes has increased from 23.4% of America's wealth to 31.2% in 2022. Net wealth is the value of a person's assets minus the cost of outstanding debts. And that increase for the top 1% came from the bottom 90% of American incomes dropping from 39.7% of net wealth in 1989 to 31% in 2022.[87] And to be specific it isn't just the poor getting poorer as the biggest part of that drop in wealth occurred between the 50th percentile up to the 90th percentile. The middle class of America's income spectrum had their share of America's wealth reduced from 35.9% down to 28.4%.

Think about that, the top 1% of America's incomes owns 31.2% of the net wealth, which is more than the bottom 90% of households owns. And this should not be surprising when you consider that the top 1% were paying a lot more in taxes in 1980 and have received the lion's share of the tax breaks since 1980, while the lower 90% received far less in tax reductions.

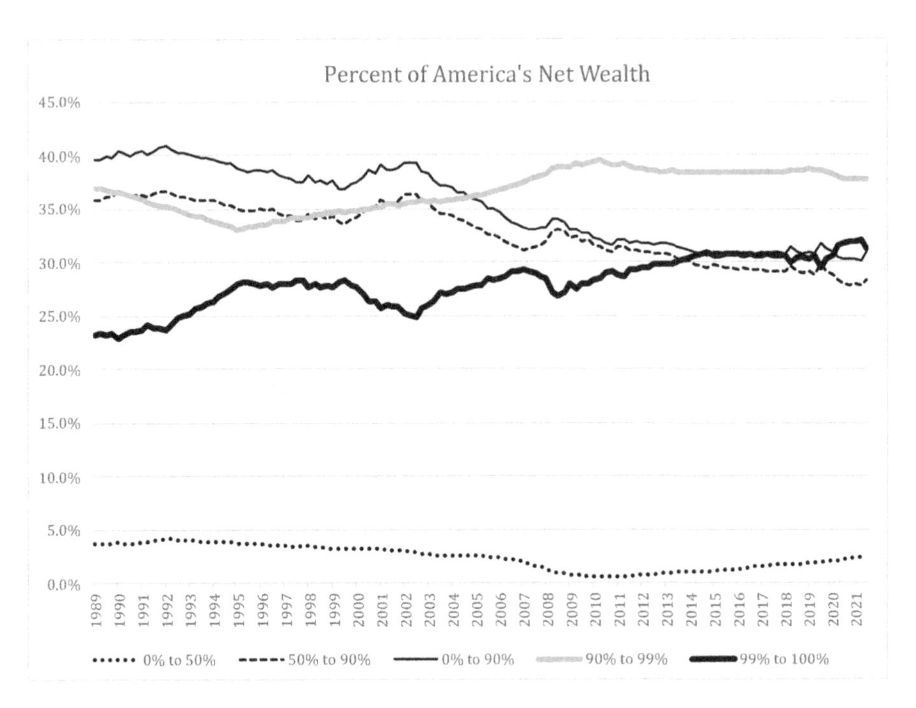

FIGURE 4. WEALTH DISTRIBUTION OVER TIME

Financial inequality is recognized as a problem in the United States by 61% of Americans, though it splits along party lines with only 41% of Republicans indicating it is a problem, while 78% of Democrats or those leaning Democrat view it as a problem.[88] Only 42% of Americans feel it should be a top federal priority.

One of the arguments that you hear against raising taxes on the wealthy is that they already pay the majority of the taxes, and based on data from the Congressional Budget Office, this is correct.[89] The 10% of America that is the wealthiest pays 54% of the federal income taxes when they have 40% of the income. That's right, the top 10% of the country makes almost half the income for the whole country. And it gets even worse when you look at wealth. The top 10% owns 69% of the country's net wealth, hence the high inequality coefficient.

But the problem with that statement is it doesn't look at the cost of living. If you were to assign, say, a basic cost of living of $4,000 per month to everyone, and then look at what we will call the "excess income," all the income above the first $4,000 per month, where does that reside? In that case, the top 10% has 58.8% of the excess income and pays 54.3% of the federal taxes. And let's be realistic, squeezing the bottom of the income spectrum for taxes can be done, but every dollar you take from them is one less dollar they will spend in the economy which helps to power it. The top 10% has a lot more excess money that can be taxed.

So, when faced with mountainous debt and financial inequality, the first thing to do is to adjust the tax structures to something much closer to where it was after World War I. Republicans during the Reagan years made a big deal out of reducing the number of tax brackets, but this only benefited the very wealthy. It's time to increase the number of brackets and raise the ones for the wealthy above 40%. If we can't balance the budget at those rates, then we need to raise them until we're running budget surpluses and paying off the national debt.

If the wealthy don't like that idea, let's look at the other options. Either raise taxes on the middle class and the poor, which increases income inequality, decreases the economic activity, and leads to civil unrest; another option is massive cuts to Social Security, Medicare, and Medicaid, which will leave the elderly and poor unable to pay bills or get healthcare; or continue to run deficits until the government defaults or the economy crashes, which will also lead to civil unrest and other economic issues. Those really aren't long-term options, so let's just agree to increase taxes on those making more than $300,000 per year.

But the really wealthy aren't paid in wages. It is the long-term capital gains tax that interests them the most. For households whose annual income is less than $500,000, the capital gains tax is 15%, while for those making over $500,000 per year it's 20%. Compare that with the income tax rate of 37% for those making over $500,000 per year, and you begin to see why most of the wealthy people make their money via capital gains as opposed to salary. For the top 1% of income earners, those making more than $750,000 per year, they make over 38% of their income in capital gains on investments. It gets even more lopsided for the top 0.1%, those making more than $1.6 million per year. They get almost 60% on average from capital gains. Even private equity fund managers also enjoy the tax break when they take their compensation as a portion of the realized gain from the investments they manage, so they pay only capital gains tax and do not pay income tax, a difference of 17% that allows them to pay almost half the taxes other people would pay.

And just like the income tax, because there are only two brackets, when you talk about raising the capital gains tax you get everyone making more than $500,000 a year interested, even when we are aiming to increase the tax burden on the multi-millionaires and billionaires. The solution is simple, increase the number of brackets, set up two or more additional brackets, with two higher capital gains taxes on those brackets, say 30% and 40%. Not only adding brackets, but also require taxes to be paid anytime the assets are transferred. If stocks are transferred from one entity to another entity, even if the same people are in both entities, make them pay a tax on the value at the time of transfer.

The next question on taxes is the corporate tax rate. With the tax cuts of 2018, it stands at 21% for companies of all sizes from small mom-and-pops to the giant international companies. Looking at the same time frame, 1920 to 2020, after World War I the corporate tax rate had been 10% and through the 1920s and 1930s, it gradually rose to about 18%. During World War II, it rose to 40% and in the years after it rose to 52% and then gradually dropped to 48% before it was cut to 34% at the end of the Reagan administration and then to 21% during the Trump administration. But the problem is that the effective tax rate has been falling since the 1950s. As corporations have become smarter about hiding their profits and using tax havens, the effective tax rate has fallen into the low teens, and horror stories of corporate giants paying zero corporate income tax are rife.

But the question has to be asked: if we have a progressive income tax where small earners pay less than big earners, why don't we have a progressive corporate tax where small corporations continued to pay 21% or maybe 20%, while big corporations have to pay 30% to 40%?

Why do this? Well, first it makes big corporations less profitable, and wouldn't that be a good thing? Big corporations have been gobbling up small companies and usually that means a nice payout for the small company owners. But what does it mean for the small company employees and the neighborhoods they were located in? There is usually a consolidation of duplicate organizations. Layoffs and closures tend to follow in the wake of these mergers, and it is rarely beneficial to anyone except the stockholders and corporate executives. Many American corporations have crossed the threshold to international corporations. The encouragement of small locally owned companies has to be considered a good thing. Small business has been a strong backbone of America and especially in small towns. The impact on American small towns has been devastating as jobs have dried up and moved elsewhere. It is time to make corporate taxes progressive in their encouragement and protection of small business.

There is a bigger issue that large, multinational corporations are moving profits to countries with low tax rates. Since 1965, US multinational corporations have been moving their reported profits to other countries. In 1965, about 5% of profit was reported in foreign countries that are considered tax havens. In 2020, it was 60%. Not only did they manage to massively increase their profit by moving money to these countries, but it was done without increasing the payroll or capital investments in those countries.

How is this done? Well, one example is by transferring intellectual property from the United States where it was developed to a foreign tax haven where it can generate profits. Google's search and advertisement technology was transferred to a subsidiary of Google known as Google Holdings located for tax purposes in Bermuda. In 2017, Google Holdings registered $22.7 billion in revenue in Bermuda, which has a corporate income tax of 0%. That means that all the other branches of Google were "paying" Google Holdings to use the intellectual property, and the profit was being taxed in the Bahamas. And it's not just Google. Apple and other companies book a lot of their revenue in Ireland. Why? Ireland corporate tax was 12.5%.

In 2021, most countries signed on to a minimum corporate tax of 15%, but that increase is not scheduled to take effect until the end of 2023, so it still remains to be seen what it does to Bermuda, Ireland, and the other tax havens that appear to have signed on.

How does the United States deal with corporations that generate revenue in the United States and then shift that money to other low-tax countries by means which are purely accounting tricks. The idea that intellectual property can be developed in the United States, sometimes with taxpayer assistance and then moved to a foreign country should be disallowed and all past moves questioned.

The basic fact is that with over $31 trillion in debt at the end of 2022 and deficits predicted in the foreseeable future, federal revenue must increase to provide a balanced budget with some ability to begin to pay down the national debt. The solution must begin with revising tax rates to provide for increased revenue and financing the Internal Revenue Service so they can do their job and start ensuring that everyone pays their taxes fairly. President Biden recently added millions to the IRS, and conservatives cried foul. While the IRS has been vilified for years, it has also systematically shrunk over the last twenty years. In 1996, they did 16.7 audits per 1,000 individual tax returns; in 2018, it was six audits per 1,000 tax returns. In 1996, IRS operating costs were 0.0897% of the GDP; in 2019, it was 0.0552%. Gross collections which are tax collections prior to refunds was also down 1.8%. While some will claim that most of the IRS filings are electronic now, over 21 million fillings in 2021 were on paper and created a massive backlog that had to be hand-entered by IRS staff.

Audits for corporations are also down from 2.28% being audited in 1996 to 0.87% in 2018. This shrinking has been done by cutting the IRS operating budget from $14 billion dollars in 2010 to $12 billion dollars in 2017. Adjusting for inflation the 2017 budget was a 24% decrease compared with the 2010 budget. A ProPublica article in 2018 detailed how these cuts have caused the number of auditors to drop by a third from 2010 to 2017[90]. There is no hope of generating more income for the federal government if the watchdog, the IRS is not funded to the point where it can ensure that people are paying their fair share. An increase in funding has already been initiated by the Biden Administration and needs to be implemented.

Not only are wealthy Americans and American corporations the beneficiary of this reduction in the IRS, but foreign companies are also avoiding taxes. In 2008, IRS auditor William Pfeil identified hundreds of foreign companies that had avoided US taxes. He estimated his group brought in $50 million until he retired in 2013 as a result of the cuts that the IRS was undergoing. He understands his group was shutdown not long after he left.

While conservatives are portraying the increased IRS funding as a threat to middle America, it really is a threat to the people not paying their taxes. Which with the massive federal deficits, is something that needs to be addressed.

Some argue that less spending is the solution to the national debt; the problem is where to make those cuts. For 2019, the legally mandated social safety nets, Social Security, Medicare, Medicaid, and Unemployment Insurance made up 62% of federal spending. Add in payments on the interest for the national debt and defense spending, and that is 77% of the budget and 99% of the federal revenue. Several of the programs and policies defined in earlier chapters will strive to reduce spending, but there is no avoiding that additional revenue is needed.

So, what do we do?

- For personal income tax, add more brackets and increase taxes on people making more than $300,000 per year. As was shown, the people making less than $100,000 are paying almost what their parents paid while the wealthy are paying a fraction.
- For capital gains taxes, add more brackets and again adjust so that people earning more pay more. We also end the ability of money managers to pay what is clearly income at capital gains taxes for them.
- For corporate taxes, add brackets and set it up where small companies pay less, and big companies pay more. And also require that corporations not transfer intellectual property from where it is developed to low tax havens there by avoiding corporate taxes.
- For the IRS, give them more funding and reevaluate the results in a few years.

31.
Supply-Side Economics

> The rich are always going to say that, you know, just give us more money and we'll go out and spend more and then it will all trickle down to the rest of you. But that has not worked the last ten years, and I hope the American public is catching on.
>
> —Warren Buffett

Warren Buffett made the above statement about trickle-down economics in 2011. Since then, the concept has been used to justify the Republican tax cuts in 2017 and is a Republican mantra to this day. The reality is that the 2017 personal income tax cuts were temporary and are already beginning to expire. Only the corporate tax cuts were permanent.

Republicans have been using the supply-side economic theory since the 1880s. Historian Heather Cox Richardson documents a June 1889 article by Andrew Carnegie published in the Republican magazine *North American Review* in which argued that income inequality was a good thing, and that the wealthy class furthered the "refinements of civilization."[91] He indicated that if the poor were given better wages, they would only squander it on necessities and "small luxuries." In 1924, Andrew Mellon, who had been appointed Secretary of the Treasury, quoted Henry Ford in his book that "high taxes on the rich do not take burdens off the poor. They put burdens on the poor." Mellon worked to reduce taxes on the rich, believing it would spark the economy. He then got to preside over the massive economic crash of 1929 after the wealthy had created a stock bubble with all the extra cash that the reduced taxes had provided.

A recent test of supply-side economics was conducted in Kansas starting in 2012. Governor Sam Brownback passed sweeping cuts to both income and corporate taxes, reducing taxes on rich Kansans from 6.45% to 4.9%. While the rich got a 1.5% tax break, poor Kansans only got a 0.5% reduction. The law reduced taxes by $231 million in the first year. Brownback promised it would create 23,000 jobs and provide rapid economic growth. It would be "like a shot of adrenaline into the heart of the Kansas economy." It wasn't. By 2017, state revenues continued to fall, and spending on roads, bridges, and education had been cut. The promised growth hadn't occurred, the Kansas economy had grown slower than that of the rest of the country. The Republican legislature faced more cuts and complaints from voters, so they rolled back the tax cuts. When Brownback vetoed the rollback, the legislature overrode the veto and reinstated the pre-2012 tax rates.

Studies have shown that tax cuts promote economic growth, but the increased revenue never makes up the amount lost by the tax cuts, so that cuts in spending are required to avoid deficits. People point to the Reagan years (1981 to 1988) as an example of supply-side success, but they miss the deficits that ballooned under Reagan. While the GDP under Reagan increased from $3.2 trillion to $5.2 trillion, when adjusted for inflation, it was only 30% growth in eight years. Respectable but not outstanding. At the same time, the national debt grew from $907 billion to $2.6 trillion, tripling in the same time frame.

There is a second theory of economics—the demand side, sometimes called *aggregate demand* or *domestic final demand*. It argues that by increasing demand, you increase the economy. And we see this when consumer spending drops, the economy takes a hit. It doesn't matter how much stuff is sitting on the department store shelves. If people aren't buying, then the economy isn't moving.

Which brings up the question: how do you stimulate the economy? Research reported in the *Chicago Booth Review* looked at which of the multiple stimulus efforts during COVID was most beneficial to the economy and came to the conclusion that it was providing money to people who would spend it on basic necessities.[92] The Republicans have been using the adage "a rising tide floats all boats" since the 1880s as a justification for giving a tax break to the wealthy. That hasn't worked, but increasing the flow of money through the economy by providing it to America's poor would raise the tide and provide an advantage to everyone.

So how do you increase the flow of money in the economy? There are two options, increasing minimum wage and providing direct checks in what is called Universal Basic Income.

32.
The Minimum Wage

Only in our dreams are we free. The rest of the time we need wages.

—Terry Pratchett

There has been a great deal of discussion about raising the federal minimum wage to $15 per hour. Bernie Sanders even recently proposed a raise to $17 per hour. And arguments on either side have been intense. Conservatives argue that it would cause businesses to fail, while progressives argue it will lift people out of poverty, and as in so many other cases, both sides are right. But we need to dig into the details to understand why they are both right and what the real answer is.

First, we need to understand that the United States is not homogenous. Each state is unique, and each state has a different economy. The average GDP was $65,297 per person in 2019 across the United States. But individual states ranged from New York at $91,102 down to Mississippi at $38,966 per person. But this also corresponds to a cost of living that is not the same across the country. Mississippi has the cheapest cost of living, while New York has the second highest, with Hawaii the most expensive state to live in.[93] So, while New York state's minimum wage is $12.50 per hour, the minimum wage for New York City is $15 and some cities nearby are $14 per hour. And while $15 per hour may be required in New York City, is it required in Mississippi where the cost of living is half of what it is in New York?

Mississippi does not have a state minimum wage law, and as such, is governed by the federal law of $7.25 per hour, which was set in July 2009. Seven states either have no minimum wage law, or in the case of Georgia and Wyoming, laws set to $5.15 per hour, far below the federal level.

State Economic Ranking by GDP $/person	Minimum Wage per hour			Cost of Living Index
	2021	2019	2011	
Top 5	$13.14	$11.42	$8.03	131.0
Top 10	$11.63	$10.19	$7.69	123.0
Top 20	$10.36	$9.33	$7.53	116.1
Bottom 20	$8.81	$8.36	$7.29	94.8
Bottom 10	$8.74	$8.28	$7.31	94.2
Bottom 5	$8.30	$7.95	$7.25	89.7

TABLE 9. MINIMUM WAGES ACROSS AMERICA

Every state in the top ten economies has raised its local state minimum wage except one, North Dakota. And North Dakota was not in the top ten in 2000. It joined the top ten based on shale oil production which has spurred the state economy and has a very low cost of living indexes at 96.5. Of the remaining nine states in the top ten, Delaware has the lowest minimum wage at $9.25/hour, but it also has the lowest cost of living of the remaining nine states at 107.7. In the bottom twenty state economies, the ones with the highest cost-of-living indexes have all raised their minimum wage. Vermont has a cost-of-living index of 116.4 (highest in the bottom twenty), and its 2021 minimum wage was $11.75 per hour.

Which came first: the high cost of living or an increased minimum wage? Kind of a chicken-egg dilemma in economics. In the bottom twenty, there are three states that may help resolve this question. Arkansas (49th in state economies), New Mexico (42nd), and Missouri (37th) have all raised their state minimum wages to $10.30 per hour (Missouri) or higher in 2020. All three states have also declined in terms of the state GDP per person since 2000. Missouri dropped eleven places from 26th to 37th overall. New Mexico dropped six places and Arkansas dropped two. Monitoring how these state economies perform over the next five to ten years could provide insight into how changing minimum wage impacts a state's economy and cost of living.

While we think of fulltime workers supporting a family when we say minimum wage, a lot of the people earning minimum wage are teenagers. A *Wall Street Journal* article from March 2021 claimed that sixteen- to nineteen-year-olds made up 37.5% of those being paid the $7.25 federal minimum wage.[94] The Department of Labor statistics don't back up the *Wall Street Journal*'s claim. In fact, they show 272,000 sixteen- to nineteen-year-olds working at or below minimum wage out of a total of 1.6 million workers being paid minimum wage or less.[95] Of the fulltime workers that are part of that 1.6 million, only 98,000 are paid minimum wage. The majority are paid less. The minimum wage becomes an issue for part-time workers and those making less than minimum wage. The way the laws are written, if you receive tips and get to keep the tips, then your employer can pay you less than the minimum wage, provided the tips make up the difference to meet minimum wage. It's also written that those under twenty years old can be paid less than minimum wage for their first ninety consecutive days of employment, so teenagers starting out don't qualify for minimum wage at first. There are also a number of employment categories that are exempted, such as babysitters, companions for the elderly, newspaper deliverers, and farm workers on small farms.

The Congressional Budget Office released a study on the impact of a $15 per hour minimum wage and indicated that while about 900,000 people would be lifted out of poverty, another 1.4 million Americans would lose their work.[96] Those losing work would be the young and less educated. The report indicates that the increase would also raise the costs for goods and services, resulting in an increase in the deficit by $54 billion over the next ten years.

There are some studies by economists that show an increase helps employees, and others that show it hurts employees. A study of restaurants in New Jersey (which raised the minimum wage) located next to Pennsylvania (which had a minimum wage of $7.25) showed no job loss in New Jersey. But another study in San Francisco showed some restaurants closed when the minimum wage was raised. Add to this the complexity of the gig economy where workers are now independent contractors and companies don't have to worry about minimum wages, healthcare benefits, or paying employment taxes, and this question becomes even bigger.

What is the objective of this whole discussion—to reduce poverty? And how do we go about it? Previously we showed that reducing poverty primarily required getting an education and a full-time job and ensuring that the opportunities exist for both the education and the jobs. Education we already discussed, so the real objective is to make sure that fulltime jobs are available and that they can help lift people out of poverty.

But what is the poverty line? The US Census sets the poverty thresholds, and for 2022, it indicated that a single person making less than $14,891 per year was living in poverty. Add two kids and the threshold is now $23,284 per year. How does that stack up against the federal minimum wage, presently set at $7.25 per hour? Assuming the person has a fulltime job, that's about 2,080 hours per year (fifty-two weeks at forty hours per week), that is $15,080 per year. A single person working at the minimum wage at a full-time job barely exceeds the poverty level. What about the single mother with two kids? She needs a wage of $11.20/hour, or she needs to work over 3,200 hours a year at $7.25/hour to exceed the poverty level. Only 13 states and the District of Columbia have a minimum wage greater than $11/hour.

But wait, that's an old fashioned—forty hours a week, fulltime job with paid holidays. Many jobs are now gig economy jobs that offer flexible hours but no benefits. Some companies even limit the workers to less than forty hours a week to keep from classifying them as "full-time" and being required to give them full benefits. Part-time workers can receive reduced benefits. If you circle back to the single mother, she has to have some form of childcare in order to work, which can eat up 10% or more of her income. And that's how we end up with people working multiple jobs to try and get out of poverty.

Now these poverty thresholds are the average for the forty-eight states that make up the mainland. Alaska and Hawaii are not included and have a higher cost of living. Even in the mainland, if you looked at Mississippi (the lowest cost of living) and New York (highest in the mainland), you get different poverty thresholds. But the big takeaway is that the minimum wage is geared at keeping a single person above the average threshold and doesn't apply to anyone who is supporting children.

But let's get back to that single mother with two kids. How would she be doing in say 1981, over forty years ago? The poverty threshold for a single adult was $4,284 ($13,794 in 2022 dollars), but for a mother with two kids, it was $6,635 ($21,365 in 2022 dollars) and the minimum wage was $3.35 per hour. And there were a lot more forty-hours-a-week jobs where she could then make $6,968 per year.

The issue is that the $3.35 minimum wage of 1981, when adjusted for inflation, would end up as $10.79 in 2022 dollars. Even the present minimum wage of $7.25 that was enacted in 2009 would end up when adjusted for inflation being about $9.86 per hour in 2022, even more now. The present federal minimum wage has not kept up with inflation, especially for those living in states with higher costs of living. So, does that mean that we need a national $15 per hour minimum wage?

It goes back to the diversity in America's cost of living across the states. In New York City and other areas that have a very high cost of living, $15 per hour is likely required, but not in Mississippi or the states that have very low costs of living. While progressives assert with confidence that raising the minimum wage to $15 per hour will help people out of poverty, the economic studies don't totally support that. The minimum wage increases in Florida, Missouri, Arkansas, and New Mexico provide perfect case studies to see how changing the minimum wage impacts employment levels and the amount of people living in poverty to determine if it really has a clear benefit. Especially since each of these states has made the increase in a different way, it should be able to narrow down the impact of changing the minimum wage. And that is what should be done before rushing into a push to make a $15-per-hour minimum wage across the nation. Doubling the minimum wage could have unintended consequences on state economies with the weakest GDP per citizen. States that can ill afford to have businesses closing, or have those business cut back hours for employees, or reduce services, or raise prices and drive up the cost of living.

The present solution, like so many things, is not a sudden change to $15, but rather a slow change over a period of years to something around $10. That may not satisfy everyone, but realize that most states with higher costs of living are already well above $10 per hour and this addresses the poorest states with the lowest cost of living that haven't had an increase since 2009.

Now the uglier question: what do you do about all those jobs excluded from the minimum wage, especially people who earn tips?

Let's just focus on restaurants as an example of how this works. The customer orders food, the waiter brings the food, the customer eats, and then in the act of paying, the customer adds a tip of 10–20%. The restaurant then doesn't have to pay the waiter minimum wage. They pay them less, and the waiter gets to keep part of the tip. Usually, the tip gets shared with bussers and other staff who don't have to receive minimum wage. So, the customer is paying the waiter's salary with the tip and reducing what the restaurant pays the waiter, which is defeating the whole reason the customer tipped in the first place. The tip is for good service; it's supposed to be an extra for the waiter and staff of the restaurant, but it has morphed into a way to supplement restaurants' payroll costs. And why is that a good thing? Because if you take that away, then food prices have to go up. Some will argue that the owner makes less money, but let's face it: most of the time we are not talking about a fancy steakhouse, we are talking about the mom-and-pop that is barely getting by. Remember San Francisco. A raise in minimum wage caused some small restaurants to go under. And while it may make people feel good to think they are sticking it to some top-dollar steakhouse owner, don't kid yourselves, they won't hesitate to raise their prices and pass this along to the customers. This will have the most impact on the mom-and-pop category, family-owned, small, and usually barely getting by. Independent restaurants with one or two locations made up 53% of the restaurants in the country, as reported by the NPD Group, a global market information company, in May 2022.[97] So, we have a simple choice: keep the system where people who get tips aren't covered fully by the minimum wage, or have the minimum wage apply to them and try and slowly change the system so that we don't put lots of restaurants out of business.

And what about retail staff and seasonal salespeople who are dependent on sales commissions? A lot of the retail stores pay the sales staff a commission on items sold, but if the item gets returned, they lose the money and if they don't sell enough, they don't get paid enough. The retail stores that do this, have taken the business risk of slow sales, and forced their employees to accept part of that risk. Bad week with no one coming in to buy stuff, the employee suffers as much as the store. Wonder why sales associates are almost fighting over a wealthy customer? Take a guess. Few stores share commissions among the whole staff, and that turns the sales floor into a competitive environment. It also means that the sales associates are working their contacts and trying to get people to come into the store. That benefits the store, but it is a risk for the employee.

A further complication is that as more retail sales have moved online, the risk to sales associates has increased. Another example of a career path that used to be secure and is now replaced by automation. As much as some people want to howl about this, it is standard practice. How many blacksmiths do you know? They used to be quite common before automobiles. There are a lot of bygone jobs that don't exist today: switchboard operator, telegraphist, milkman, and others. Changes in society and technology have always created a shifting employment landscape that has rewarded some and crushed others. With fast food restaurants, grocery stores, and warehouse stores installing self-checkouts and order kiosks, cashiers' jobs are being replaced with automation. As much as we want to ensure that people have good jobs that can keep people out of poverty and to hold back the tide of automation to protect those jobs, it runs contrary to the fact that we need to respect that the employment landscape is going to change.

There are times when you need to get the participants to sit around a table and discuss the issue, and this seems like one of them. As much as I'd like to see a system where we convert waiters and other people forced to live on tips and commissions and protect the jobs of cashiers, I think this needs to be discussed, as it has a direct impact on prices and profitability. That discussion needs to be about the other jobs presently excluded from minimum wage and what needs to be done about them. There is a gray line for government between preventing the exploitation of workers and interfering with the economy.

Now, the bigger issue is the gig economy and its impact on workers. The definition for a gig economy is "a labor market characterized by the prevalence of short-term contracts or freelance work as opposed to permanent jobs." Now there have always been consultants and freelancers who have worked on short assignments and moved around. The very term *freelancer* originated in the early 1800s to refer to a medieval lance ("knight") not sworn to any lord's service, i.e., a mercenary. And these people have always been vulnerable to getting laid off, but the problem is that this employment has morphed in recent years. For people with technical skills and a desire to travel and change employers, being a consultant/freelancer makes sense. The problem is that Lyft, Uber, DoorDash, GrubHub, and a multitude of other online service providers have carried the model to the extreme, where all drivers and front-line workers are contractors. It reduces their overhead costs and pushes payroll taxes and health insurance on to the contractors.

Estimates are that 36% of American workers are part of the gig economy. They tend to work fewer hours (twenty-five hours a week versus forty hours for a full-time employee), make less money, and have less benefits. For some, it is a choice to work less and have more flexibility, but for others it is the only job they can find as so many other blue-collar jobs have faded away.

The gig economy has a huge variety to it, and a lot of people are quite happy working this way. But the question is what to do about people who work for apps like Uber, Lyft, and DoorDash who aren't making enough to stay out of poverty and don't have health or retirement benefits. This is especially true where the drivers don't own their cars. There are a number of "non-driving partners" who own the cars, but then have other people drive them. Uber doesn't pay the drivers; they send the money to the car owner, and then they handle paying the drivers. If you think that is open for a lot of abuse by these car-owning middlemen, it is. Uber will even lease a car to drivers, and a number of Uber drivers in the Washington, DC, area indicated that this led to a debt trap where the driver ended up making around $5 per hour.[98] Again, there are a lot of scenarios and outcomes here, but the question is what to do about the people who are falling through the cracks and the companies who are trying to widen those cracks? Some will think people are accountable for their own mistakes. But we need to understand if the apps have gamed the system to create an unwinnable situation at the workers' expense. At coal mines in the middle of nowhere, workers had to pay for their lodging and all supplies from the company store. The 1947 song "Sixteen Tons" had the famous lyrics:

> You load sixteen tons, what do you get?
> Another day older and deeper in debt.
> St. Peter, don't you call me cause I can't go,
> I owe my soul to the company store.

The concept of a company store is a pattern where the workers are kept dependent on the company, and the explosion of the gig economy has a lot of opportunity for these companies to take advantage of all their contractors without many repercussions.

In California, an attempt was made to reclassify app-based transportation and delivery drivers from contractors to employees, which would have made them eligible for benefits and made the app companies responsible for part of the payroll taxes and several other issues. The internet companies pushed back by getting Proposition 22 on the November 2020 ballot that exempted them from this requirement, and they spent an estimated

$203 million to support the effort to get it passed. The proposition was passed by California voters by 58.6% for and 41.4% against. Now, this proposition wasn't all bad for the drivers; it had a few benefits for them but did keep them as independent contractors, and that may be the solution. Establishing a third type of worker besides the classic employee or independent contractor reflects and protects gig economy workers, while still respecting their independence and flexibility and providing some benefit for app companies.

CALIFORNIA PROP 22 PROVIDED THE FOLLOWING CHANGES:

- payments for the difference between a worker's net earnings, excluding tips, and a net earnings floor based on 120% of the minimum wage applied to a driver's engaged time (the time between accepting a service request and completing the request) and $0.30, adjusted for inflation after 2021, per engaged mile;
- limiting app-based drivers from working more than twelve hours during a 24-hour period, unless the driver has been logged off for an uninterrupted six hours;
- for drivers who average at least twenty-five hours per week of engaged time in a quarter, require companies to provide healthcare subsidies equal to 82% the average California Covered (CC) premium for each month;
- for drivers who average between fifteen and twenty-five hours per week of engaged time during a quarter, require companies to provide healthcare subsidies equal to 41% the average CC premium for each month;
- require companies to provide or make available occupational accident insurance to cover at least $1 million in medical expenses and lost income resulting from injuries suffered while a driver

was online (defined as when the driver is using the app and can receive service requests) but not engaged in personal activities;

- require the occupational accident insurance to provide disability payments of 66 percent of a driver's average weekly earnings during the previous four weeks before the injuries suffered (while the driver was online but not engaged in personal activities) for upwards of 104 weeks (about 2 years);
- require companies to provide or make available accidental death insurance for the benefit of a driver's spouse, children, or other dependents when the driver dies while using the app;

This could be used as an example to begin to form a federal guideline for gig economy workers.

The Minimum Wage Policy

- To not raise the minimum wage to $15 per hour, but rather to have Congress raise the federal minimum wage to $10/hour over a span of five years.
- To also review California Proposition 22 as a basis for defining a set of minimum requirements for gig economy workers.

33.
Universal Basic Income

The fee to cover the average cost of incarceration for federal inmates was $34,704.12 ($94.82 per day) in FY 2016.

— Federal Bureau of Prisons

Universal Basic Income (UBI) is an idea that is being advanced by several individuals and groups right now. The idea is every adult gets $1,000 or some other amount per month from the government to help provide financial stability. For some Americans this isn't much, but for households that make less than $10,000 per year (~6%), this could be a major life improvement. Consider that 38.4% of households make less than $50,000 per year, and this would be a real improvement to their lives.[99] The question is what would be the tangible benefit? And if you wanted to do it, how do you implement it and pay for it?

Studies have shown improvements for poor people provided a monthly payment, but these have mainly been small-scale tests in Western countries, though there have been some larger tests in third-world countries. A nationwide test has not been conducted on the scale envisioned. While the existing tests have shown benefits for poor people's personal outcomes, this may be offset if UBI causes inflation or economic impacts based on taxation. Some economists show that lower-income families are very likely to spend it on necessities, thereby providing an economic stimulus to the overall country. Some have argued this could cause inflation, others that it could spur economic growth, increased jobs, and very positive outcomes.

The state of Alaska has been giving each Alaskan citizen a check since 1982 from the Alaska Permanent Fund, which is a state-owned investment fund financed by oil revenues. While the check amount varies, it presently averages about $2,000 per year per person ($167/month). Not a lot of money, but Alaska doesn't report the extreme poverty that most other states do. Now whether that is due to the harsh winters, smaller population, or the dividend check is not clear. A 2019 study by economists showed that the dividend check wasn't negatively impacting employment within the state and has provided some positive results.[100]

The Eastern Band of Cherokee Indians have been providing a casino dividend to all its tribal members since 1997 and again, there have been no repercussions to employment, and they have seen positive impacts in education, mental health, crime, and drug addiction. The dividend averages between $4,000 to $6,000 per year ($333 to $500/month).

Some small trials have been conducted elsewhere in the United States and Canada, and the results have been positive, but the government-sponsored ones have a tendency to get cancelled early when a conservative government takes over. The conservative view is that it is a bad idea to give away free money to the poor.

The scientific consensus of UBI tests is that UBI decreases poverty. The studies also show that UBI for poor people tends to go towards food, rent, and utilities, the necessities, but does spark increased economic activity as a result. The studies do not show a decrease in employment, except where those hours are channeled into caregiving, either for children, incapacitated relatives, or the elderly. Another point is that these studies weren't "universal"; they were focused on poor and low-income families.

While there has been a lot of theorizing that "giving away" money to the poor can't be a good thing, the basic result for the studies has been that UBI does have a positive outcome for people in lower incomes. The problem is that many look unfavorably on a giveaway but have no issue paying for prisons or other things that seem to cost more per person than UBI would. The question seems to be: are we ready to try something different, giving money to the poor to try and end poverty? But we need to accept that this will not end poverty, it may provide the steppingstone for more people to get out of poverty. And that is the objective. Too many of our problems trace back to poverty, and too much of the nation's poverty stems from the racism that was prevalent before the 1970s. While some will push back on this idea, the proposal is to attempt to provide a UBI for the very poor and after ten years, to step back and evaluate it. Has crime decreased? Has spending on prisons decreased? Has the general welfare of Americans improved? If so, then we need to continue and think about expanding the program. If not, then we can consider shutting it down.

This brings up the next set of questions. How do you implement it, what does it cost, and how do you pay for it?

Rolling out a program where every American adult gets $1,000 per month would require a herculean effort, so the solution is to start small and gradually ratchet it up until we are getting less return on investment. Think of it this way: UBI helps the poorest people by providing them with money to help meet their basic needs. Once we start providing it to middle-income families, the benefit goes down. Not that those families won't appreciate the extra money, but they wouldn't be spending it on necessities. In America, 37 million people are living in poverty, and of those, 26 million are adults. Those 26 million are where we will get the most positive return. The way to implement this would be that everyone would have to submit an IRS tax return and have an account set up for direct deposit. That is going to create a hurdle for homeless people, but hopefully non-profits and charities could step in and provide a way for the homeless to gain access to the money. And there will be some who will not want to participate and won't share their information, and this is fine. If someone chooses not to participate, that is their decision, however misguided it might be. We need to respect that people may choose to sink and not swim. It makes no sense to force them if they won't participate. So, this will be voluntary participation.

Also, it should only apply to US citizens. If you have come to America and have a visa or legal permanent residence, you don't qualify. That may seem unfair, but it limits the cost to the government and provides no incentive for people to immigrate, which, as we discussed in the chapter on immigration, is a problem. Once someone gains citizenship, they are eligible.

If someone comes to America, they need to take care of themselves and not depend on government handouts.

And we don't give it to everybody. The key is to start with the poorest people and work our way up the economic ladder as funds become available. We don't borrow money to do this, we gather money the first year, and based on how much was gathered we distribute it the second year based on some simple criteria. So, it would not be universal basic income, it would be just basic income (BI). And it would only apply to people who filed an IRS tax return the previous year (how we determine income status), have a bank account we can deposit the monthly money into, are US citizens, and not incarcerated. If you are in jail, society is already paying for your room and board, so you don't get BI.

So how do we decide based on income who gets the additional money? Set an income threshold where payments begin to taper off and a point where it stops. Let's say initially it was \$12,000 to start tapering off and \$36,000 per year per adult where it stops. If you make over \$36,000 per year, you don't qualify for BI, if you make less than \$12,000 per year you get the full \$1,000 per month. Make between \$12,000 and \$36,000, you get a sliding scale. Someone making \$24,000 a year would get an additional \$6,000 per year, boosting them to \$30,000. If we choose to implement BI, this is how it would start. The \$12,000 per year income where we start to taper and the threshold of \$36,000 per year could be adjusted up or down as the benefits were evaluated and how much money was available to increase the program.

So, how much money is needed?

Per the 2020 census, there are 255 million American adults (eighteen and over). A stipend of \$1,000 per month for every American adult is over \$3 trillion. About 60% of Americans make more than \$50,000 per year, so we can assume that about a third of that will be returned in the form of taxes, but that still leaves something around \$2 trillion to come up with the first year. What if progressives declare this to be non-taxable? Well, then, you're back trying to figure out how to come up with \$3 trillion. That's \$3 trillion extra when the federal budget ran up $800 billion deficits in 2018 and 2019, which were very good economic years. Let's be honest, $3 trillion to $2 trillion is too much to spend, especially when 60% of the country doesn't really need it. The whole idea is to try and reduce poverty, and since the program will just be starting, the best is to focus on those struggling.

So, if we start with just the people living in poverty, that's about 26 million, so we are looking for $300 to $400 billion a year to kick off the program with just the poor. Once the program gets started and additional funds can be found, then you can increase the income threshold and try and increase the BI coverage until the lower 40% of Americans are covered. That's $1.2 trillion—still a lot, but more manageable than $3 trillion. That is our target window. To provide a BI that starts by covering the lower 20% of Americans and protects them from the ravages of poverty and can then be increased later to cover the next 20% of Americans.

Some talk about doing it for everyone. But really, where is the benefit for the upper 60%, a sort of egalitarian inclusiveness that may seem principled, but makes this program harder to start and run? Let's focus on the people who need it.

And where is the money coming from? One idea that Andrew Yang suggested in his book *Forward* was to implement a Value Added Tax (VAT), much like European countries do, but that will cause the price of all goods to rise. Think of something akin to a national sales tax, which could reduce the benefit for poor people, but it also might reduce America's hyper-consumerism, a benefit for energy usage and trash generation. A problem is that many areas in the United States already have local sales tax, so this would be on top of the local taxes. Again, these ideas haven't been tested on a large scale within the United States and extrapolating European results is a risk.

There have been lots of other ideas for funding options—increased corporate taxes, wealth taxes, national sales tax, and national real estate taxes. They all have their pluses and minuses, and many encroach on areas where local and state taxes are generated and could create overlaps and issues.

Another idea for funding BI is a negative interest rate, which is sometimes called a demurrage. It would be a fee to all bank accounts, including savings accounts, retirement accounts, stock accounts, etc. Essentially a charge to all "hoarded" money, money not active in the economy to help fund the BI. This may seem unfair, but most investment brokers charge their customers a demurrage for maintaining the investments, and it is usually not even noticed by the investors when the investments are performing well. If you are looking for a tax on the wealthy, this just might be it. Because not only could it be applied to accounts held by individuals, but also by any held by trusts or companies. With a lot of individuals moving their assets into trusts and paper companies to avoid taxes, a demurrage might be the solution.

A very small demurrage of 0.25% could be applied at first to start the program and then adjusted as necessary. This would be a new form of tax not spelled out in the Constitution and would require an amendment. The Sixteenth Amendment was passed to allow income tax, as the Supreme Court had ruled that income tax was not allowed under the Constitution. A similar amendment would be needed to allow a demurrage or even a VAT as they are both new forms of taxation. The amendment would spell out the demurrage, a starting rate of 0.25% and restrict it not to exceed 1% and define its use for a Basic Income with the provision that after ten years, the Congress could set it to a 0% tax or raise it as required. This would allow a necessary review and decision at the ten-year point to decide if it was working.

The Federal Reserve in 2021 estimated that just the net worth of households and non-profits in the United States was $141.7 trillion dollars, which with a 0.25% demurrage would generate over $350 billion.[101] If you had a million dollars of investments and money sitting in savings accounts, that would be $2,500 per year in charges. Seem outrageous? Most investments are increasing by 3% to 12% per year. Think about it, can we not spare 0.25% to end poverty?

Once a BI system is up and running, we should dial back unemployment insurance (income security is ~$300 billion in the federal budget in non-COVID years) and Medicaid (~$400 billion) as more people qualified for low-cost health insurance which would not only help to pay for increasing BI, but also reduce the federal deficit. Also, as incarceration costs government more than the BI, we can also expect to eventually see savings in that area as well. This could allow the program to be expanded to cover more Americans by income and maybe eventually to cover the bottom 40%.

It would be a massive experiment, but the advantages could be a reduction in poverty which is the major cause of crime, poor health outcomes, and the racial and social tensions that are at almost crisis levels in America.

The Universal Basic Income Policy

- Pass an amendment to be sent to the states for approval that would establish a demurrage tax that was initially 0.25% on all savings and investment accounts that are held by citizens, legal permanent residents, trusts owned by citizens, or corporations operating in the United States.
- The funds from the taxes would begin a basic income that paid all US citizens registered with the IRS and showing an income less than an amount defined.

34.
Making it in America

The best customer of American industry is the well-paid worker.

—Franklin D. Roosevelt

While FDR, in the quote above, was referencing labor unions, it is true that well paid American workers are the customers of American industry. A frequent solution to supporting the American economy is a return to manufacturing jobs, but what has been done to increase American manufacturing? Before you can think about that, you need to understand how manufacturing jobs have changed over the years. The number of manufacturing jobs rose from 10 million to 16 million during World War II, and in the following years, it reached a high of about 19 million manufacturing jobs during the 1970s with manufacturing averaging between 17 million and 19 million jobs from 1965 to 2000. Between 2000 and 2007, manufacturing jobs declined to 13.8 million, a loss of 3.6 million jobs. Manufacturing jobs have continued to decline to 12.2 million in 2020. But when you look at this as a percent of the US population, the decline is even more drastic. In 1970, manufacturing jobs were held by 8.8% of the population, in 2020, it was down to 3.7% of the population. To find a possible reason for the decline, we need to look at the trade deficit for manufactured goods.

The trade deficit reflects the difference between the number of manufactured goods exported by the United States and the number of goods imported. For a long time, Americans have imported more manufactured goods than they have exported. Between 1989 and 1997, the difference between imported goods and exported goods was less than $131 billion, but by 2000, the US trade deficit had risen to almost $300 billion annually. In 2006, it had almost doubled to $558.5 billion per year. This reflects the transition of companies moving operations overseas to take advantage of low-cost workers and the increase in manufactured goods sourced from overseas. These decisions were made by individual companies driven by their bottom line as they searched for more cost-effective ways to make products, but it had an overall impact on the United States. However, we can't just hold the companies accountable; these decisions were also supported by an American society always ready to buy a cheaper product even if it came from abroad. Even today, you can easily find cheap products from several foreign countries sitting on American store shelves.

A relevant example is solar cells. Solar cells are going to be extremely important going into a greener energy economy, and in 2019, 71% of all solar cells were manufactured in China, only 3% were manufactured in the United States. In 2000, US production was over 30% of the solar cell market, and Japan had almost 40% of the market. The market was much smaller at that time, but US manufacturing has not kept up with the flood of solar cells from China. Japan and the United States are now less than 5% of the market together. People may point to Chinese subsidies that keep the exports cheap compared to elsewhere, but that misses the point that the United States was over 30% of the market and did not keep up. President Trump imposed tariffs on Chinese solar cells and that makes sense because it makes American manufacturers more competitive for solar projects.

In the past, the government has imposed tariffs to raise the prices of foreign products to help American products compete. But the problem now is that in several industries, there are no American products to compete. Tariffs are an added tax by the federal government that directly increase the cost of a foreign product, affecting the consumer's price. This price increase would make an American product more competitive, but it doesn't help much when there's no American product to compete. Another more recent solution is to offer tax incentives and subsidies to companies (mainly foreign) to relocate factories back to America. Tax incentives and subsidies are paid by the taxpayers with the idea that the increased jobs and economic impact will offset these giveaways. This has not happened in all cases. Some factories failed to provide the promised jobs, or when the tax incentives ran out, they just closed down. The reality is both solutions are affecting the costs to American consumers either in the product prices by tariffs or taxes when trying to entice companies to relocate. The result is by purchasing foreign manufactured products, we have lost American manufacturing jobs, which provided good jobs to Americans and did not require a college degree. Essentially, those jobs formed our middle class. We have replaced them with lower-wage service jobs, worsened the income inequality, and harmed the economies in the Rust Belt states. Is American society ready to pay more for products to support American manufacturing? How do you determine that society is willing to pay, and what do you do about it?

The first solution is to allow American buyers to support manufacturing directly by voting with their pocketbooks. That means requiring that packaging and online descriptions identify not only where the item was made, but who owns the manufacturer and where the material comes from to allow easy recognition. In other countries, large national flags depicting the source of a product are put on the front of the product, whereas presently in the United States, you need to search for the "made in" text that can be hidden. And when purchasing products online, suppliers don't indicate the country of origin. Another issue is that sometimes this represents where the final assembly was conducted, but not where the product was made, and it doesn't always correspond to the ownership of the company. Presently, the only manufacturer of children's bikes in the United States proudly indicates they are "assembled in the USA," but all the parts are made in China and shipped over.

The identification needs to be clear so people can choose to buy American. And online sales should reflect this as well. It should use flags to identify the manufacturer, the location of the source material, and the final ownership, which cannot be based on the direct manufacturer. These can be owned by companies located in other countries. The idea is to identify where the profit is directed. The law can ask, not require, that online marketplaces offer an American alternative if one exists on the website. This is specifically directed at Amazon, which routinely offers alternatives to buyers. If they could offer any from American manufacturers with the US flag displayed on the option, then American buyers would know their options.

A simple way to monitor this is after the major online suppliers have added the identification of country of origin to their US-facing sales fronts, they can track and report whether more Americans buy American products. This would be a significant vote from American society that it wants to protect and support American manufacturing. A simple report from the Department of Commerce in coordination with Amazon and other online suppliers could indicate the change in purchase preference.

The second solution is voting with the government's pocketbook. It is a common practice for countries to require that industrial projects performed in their country buy or contain a minimum amount of local content, no matter the cost. At present, the federal government has the Buy American Act (1933) that applies to most government purchases, and the Buy America Act (1982), which only applies to mass transit projects, both of which require the government to purchase things from US manufacturers unless there is not an American product available, or the US product costs are 25% more than the foreign products. There is a provision that this does not apply to countries where we have a reciprocal trade agreement. The president may also waive this requirement.

If anything, the pandemic should have taught us that relying on foreign products for critical personal protective equipment (PPE) in the medical industry is a risk in the event of any future crisis requiring medical PPE. There needs to be an adjustment of the Buy American Act to encourage the development of American suppliers for medical PPE and other technology. We should raise the 25% threshold to 30% or even 35% in the case of medical PPE, so American products can qualify for preferential purchase by the Buy American Act and be available locally in an emergency.

The third solution is tariffs. According to the Office of the United States Trade Representative, "one half of all industrial goods imports enter the United States duty free."[102] That may seem unfair, but a large part of this is the reciprocal trade treaties with close allies and favored partners, where we have dropped tariffs from their products in exchange for dropping tariffs from ours. This is part of our foreign policy, so we need to keep this. The problem with our trade imbalance isn't with close allies and favored partners; the big problems tend to be with China.

China has been working diligently to develop its economy, to increase its technical sophistication, and to increase its influence within its region and further afield. But the problem is how they have been doing this. There is evidence of state-sponsored hackers as well as the use of strong-arm tactics where Chinese authorities demand that companies hand over technology as the cost of doing business.

There can be no doubt that China is an authoritarian regime who is looking to increase its power. The Chinese authorities' treatment of the Uighurs in the far east of China, the arrests of Hong Kong democracy advocates, and the constant pressure and demands that Taiwan is merely a breakaway Chinese province rather than a sovereign state are three of the clearest examples that show how Chinese and American values do not align. Add to that China's efforts to build manmade militarized islands to extend their control of the South China Sea, areas that are closer to Vietnam, Malaysia, and the Philippines than China. And yet, they are our third-largest trading partner based on the number of exports to China. In 2019, they bought from the United States $106.4 billion of material, mainly electrical machinery ($14 billion), machinery ($13 billion), aircraft ($10 billion), optical and medical instruments ($9.7 billion), and vehicles ($9.1 billion). They also purchased agricultural products totaling $14 billion in 2019, including soybeans ($8 billion), pork and pork products ($1.3 billion), cotton ($706 million), tree nuts ($606 million), and hides and skins ($412 million).

At the same time, China is the largest importer to the United States. We purchased $451.7 billion from China in 2019. That is less than what was purchased in 2018 (down $87.6 billion or 16.2%), but more than twice what we purchased in 2009. Chinese imports are 18.1% of overall US imports in 2019.[103]

So, they are the major supplier to the United States, and they buy a lot of US goods, but they are not our friend nor are they aligned with our interests. The tariff war initiated by President Trump has been the likely cause of a drop in US imports from China, and that is not necessarily a bad thing. US policy on tariffs needs to be about two things: helping and protecting US industry and a tool of foreign policy to show displeasure when a country is not behaving correctly, both in terms of human rights and international relationships. China's technology transfer and other issues should not be rewarded with returning tariffs to lower levels until they take action that can be verified, but due to international relations, it is hard to make a case for using tariffs just to revitalize American manufacturing without impacting relations or the world's view of America, which is actually important. Tariffs are a sensitive issue and can quickly escalate into controversy.

The solution for tariffs is to implement a consumer protection fund to allow consumers to sue foreign companies and for the Bureau of Consumer Protection to identify companies not complying with labeling or design laws and to set up the fund to automatically use tariffs against countries not in compliance. That will cause manufacturers of cheap products to hopefully change their ways and increase costs or run afoul of tariffs. That way, a country couldn't blame US bias for tariffs, since we would have evidence of poor-quality products. Then the Unites States could continue to use tariffs as foreign policy tools against places that steal technology, violate human rights, or harm allies or America, but the tariffs would be divorced from an effort to protect American manufacturing.

THE MANUFACTURING POLICY

- First, require that all products sold in America, both online and in person, are clearly marked with identification of the ownership, manufacturing, and source of the material, so consumers know where their money is going.
- Second, raise the threshold for the Buy American Act to where American products are more than 30% more expensive before foreign products can be considered in place of American products.
- Third, continue to use tariffs in a foreign relations model as a consequence of a nation's behavior, but monitor the implementation of the rules under the Bureau of Consumer Protection as discussed in the chapter on tort reform, which may require specific tariffs on countries not adhering to the rules.

35.
Product Design Life

Quality is never an accident. It is always the result of intelligent effort. There is hardly anything in the world that some man cannot make a little worse and sell a little cheaper, and the people who consider price only are this man's lawful prey.

—John Ruskin

One of America's major problems is recycling and landfills, which will be addressed in the next chapter. But one item to deal with is the short life of many products. Every product has a design life, an expected time before a product wears out and must be repaired or replaced. Repairing has become nonexistent due to how cheap some products have become. Who is going to spend $20 to fix a $40 appliance when it's five years old? Not many people—especially when finding a repairman is hard and time-consuming. There is also the question of not knowing if the repairman is competent enough to repair the product.

The design life of a product is known to manufacturers. It is something that is usually set as an objective part of the design effort. Many manufacturers confirm the product design life through testing before the product is released to the public. But, at the moment, there is no requirement for the manufacturer to identify or provide the design life for the product. When consumers are faced with multiple products and must choose, all that is provided is the price. But what if the design life also had to be provided?

And then there is designed obsolescence, the design of a product that is expected to fail at some point. My own personal experience of that is best explained with a story about an electric kettle. This simple electric kettle had worked perfectly for four years and then stopped one day. The displays on the base still lit up, you could hear the relay in the base closing to put electricity to the kettle to heat up and boil water, but nothing happened. Anyone else would have tossed it in the trash and bought another one, but please remember, I'm an engineer.

Ten minutes later and using tools I keep in the garage, I had the kettle disassembled and was staring at an electric fuse that cost pennies and had been installed next to the electric heating element. The fuse was separated from the heater by a thin wafer of silvery coated material to reduce the heat of the heating element reaching the electric fuse. The fuse performs the function of breaking when a high electric current passes through it, and this is done by an element in the fuse that heats up and breaks when an excessive current causes it to heat up more than a normal electric current. But heat up and cool down that fuse repeatedly, and the electric current needed to break decreases. After four years of heating water for coffee and tea, the fuse broke without any high current event. Now, the circuit breakers of the house already provides protection to the electric circuits, so the fuse inside the kettle was redundant to the existing house electrical circuit breakers. So, I removed the fuse, reconnected the wires, and the kettle has worked perfectly ever since.

The point of this story is that manufacturers are building products with a short design life, so it fails at some point, and they hope the consumer buys a new product from the same manufacturer. There are lots of examples of this, the most infamous being the Phoebus Cartel, where lightbulb manufacturers got together in 1925 and formed a cartel with the purpose of limiting the design life of lightbulbs to 1,000 hours of use. Prior to this, lightbulbs were lasting for 2,500 hours and longer. There is a lightbulb that was first turned on in 1901, nicknamed the Centennial Light that hangs in a fire station in Livermore, California. It was originally a 60-watt lightbulb, but now it is emitting the same amount of light as a 4-watt nightlight. They still keep it lit, and it is now in the *Guinness Book of World Records*. Some defenders of the Phoebus Cartel had argued that it was really about making more efficient lightbulbs, since shorter-life bulbs produce more light with less electricity and generate less heat. The idea was to avoid letting them deteriorate, just as the Centennial Bulb presently only produces the equivalent of 4 watts of light, after over a million hours of service.

There are also examples of old refrigerators and other appliances built in our grandparents' time that still work, just like the Centennial light. If we are going to discuss issues with landfills and climate change, then one of the subjects that must be addressed is the low quality of some products that need continuous replacement. Using materials unnecessarily and adding to landfills runs contrary to the goals of environmental sustainability. And this doesn't just apply to appliances; clothing, cars, personal electronics, and even the construction of our houses have this issue.

Manufacturers should consider how long products are designed to last under normal usage. Consumer abuse of the products or not following the recommended guidelines can easily shorten the design life. So, requiring manufacturers to supply and clearly indicate the design life is not a guarantee that all products will last that long. The reason is that product failures follow a predictable trend. There are a lot of failures in the very early life of the product use, and these are referred to as the "infant mortality" of the product. These failures represent products where manufacturing errors cause products to fail early, and this is addressed by the manufacturers' standard warranty, which is usually a year with most appliances. During the product life after the infant mortality phase but prior to the design life being reached, there is usually a very small number of products that fail. But once the design life is reached, then large numbers of the product begin to fail. In industrial designs, that design life is known, and maintenance/replacement is always planned to occur before the design life is reached. Engineers call this the "bathtub curve" and can use it to reasonably predict when a product will fail.

Consumers don't have the advantage of industrial engineers; they don't know what the design life is, which is the whole point of getting manufacturers to disclose it. Never mind that a one-year warranty is offered; no one buys a kitchen appliance with the plan to replace it in a year. Consumers need to know what that design life is.

And there needs to be a mechanism to penalize manufacturers who provide false design life information or fail to meet their published design life. But it also needs to take into account the reasonable number of failures during the design life and the abuse that consumers sometimes put products through, i.e., if your child dumps your phone in the toilet, that's not a design life issue. By the way there is usually a moisture monitor in most modern phones to detect when it has been exposed to water or high levels of humidity.

So how to do this without being unfair to manufacturers or leaving consumers without recourse? The proposed solution is that a manufacturer should register a product with the Bureau of Consumer Protection, indicating the manufacturer's location (what flag/label goes on the front and online, see previous chapter), the design life for the product, the product packaging, the product's ability to be recycled (this is discussed in the next chapter), and how many of the products have been manufactured in or shipped to the United States (which would need to be updated continuously by US manufacturers or by tracking the imports). The Bureau then opens a file on the product. If people report that the product fails earlier than the design life and it is not covered by the warranty (either because the warranty had expired or the manufacturer refused the warranty), consumers can apply online or via letter to indicate their product failed. If 10% of the products fail prior to design life, that triggers an investigation. If the company is found to have supplied products that didn't meet their design life, then they must recall or refund the consumers. Foreign manufacturers that refuse to comply then get addressed under the fund created under the chapter on tort reform.

Now, there needs to be an exception for very small manufacturers, especially those making hand-crafted products. It is recommended that any manufacturer with less than a million dollars in sales should be exempt from this. This can be adjusted based on a review by the agency. There also needs to be monitoring to ensure that manufacturers, especially foreign ones, don't try and send products through multiple front companies to stay below the threshold or to change their names and restart when they run afoul of providing bad products.

36.
American Trash

We can't have landfills forever and we can't ask others to accept our trash.

—JAIME LERNER

A major problem facing America is the massive amount of trash we put into landfills and the lack of effective recycling. In 2018, the United States generated over 292 million tons of municipal solid waste, and that doesn't include construction and demolition debris or non-hazardous industrial waste. In the 1960s, it was less than 100 million tons. In the last sixty years, we have tripled our garbage generation while our population has only doubled. The problem is Americans make more garbage per person than any other country, and the bulk of it ends up in landfills. In 2018, half of the municipal solid waste ended up in landfills. That is an improvement over the 1960s when 95% of waste ended up in landfills, but the 146 million tons sent to landfills in 2018 is an issue.

There was also a lot of encouragement to the public to do more to reduce what goes into landfills. Composting, recycling, and combustion with energy recovery increased between the 1980s and 2005 with more awareness. But the increase since 2005 has been marginal, which indicates that relying just on public encouragement has plateaued, and it's time for new measures both for waste and recycling.

The first item to tackle is garbage. And the bulk of America's garbage falls into the following categories:

- Containers and packaging, 29.7%
- Food waste, 21.6%
- Durable goods, 20.4%

- Nondurable goods, 20.2%
- Yard trimmings, 12.1%
- Other, 1.5%

Food waste splits into meat products, which can't be composted easily, and bread, fruit, and vegetable waste that can be. Yard trimmings can also be composted. Containers and packaging are generally cardboard products, which can be recycled, versus the plastic wrapping and Styrofoam that present a challenge to recycling. Nondurable goods are newspapers, clothing, and similar items. Durable goods are appliances, electronics, and other items that are expected to last for more than three years.

The problem with garbage is how do you reduce it? In the previous chapter, we proposed pushing manufacturers to make products that last longer or at least tell us how long they will last. This would better inform consumers, but it would also reduce the amount of durable goods going into trash if people don't buy short-lived products. Another way to reduce durable goods is to make repairs more accessible without requiring people to have a degree in engineering. This works two ways: make it easier for consumers to repair products, and make repair centers more available. But how do you encourage making the repair of products more likely? By making throwing them away more expensive.

Making garbage more expensive would impact the manufacturers that, through their packaging and containers, create almost 30% of the garbage, and the consumers who set the bags of garbage out on the curb. At present, neither one suffers a penalty for creating more garbage, and as a result, neither has an incentive to create less. We can appeal to both groups to reduce garbage, and much has been done on that front, but let's be honest: for every consumer who diligently recycles and composts to put out a small bag of garbage every week for the garbage truck, there are many others who put out several bags. And that's the problem. From a packaging standpoint, lots of people are buying individually packaged food items so they can quickly pack lunches or take food with them, and that consumer preference is driving the manufacturers to provide individually packaged items of food.

Most metropolitan garbage collection is funded through fees or taxes that are hidden and distributed equally to all households; that way, there's no financial incentive for consumers to reduce their household waste, and when you lack financial impact for bad decisions, you can't expect change.

So, what is the solution? This goes back to a premise of this book: bad decisions need consequences, and good decisions need rewards. In this case, the person who is recycling and trying to produce less garbage needs to pay less, and the person producing excessive garbage needs to pay more. The way to do this is to weigh the garbage. On collection days, the garbage truck weighs the waste from each household and sends them a bill or adjusts their tax or fee, whatever the municipality is using. That way, less recycling and composting has a direct impact on the consumer. They get a clear indication of the cost. This requires that garbage trucks be equipped with scales, and sanitation workers will require more time on their routes, which will require more trucks and more workers, but this is the solution that turns the issue of American trash from a vague concern into something that every consumer feels directly.

On the manufacturing side, when they register a product's design life, they also need to register how much packaging the product requires and whether it is recyclable, beyond printing the little triangle symbol on it, it needs to be clearly marked so consumers know how to recycle it. The amount of non-recyclable packaging used and how non-recyclable the product is would result in a tax on the item. Call it a pre-landfill cost for the item and its packaging. The company pays that tax on all items made or imported to the United States, whether they are sold or not. Why? Because as items change hands through the supply chain, some could get lost or donated, and tracking the tax becomes much more complicated. By making the manufacturer or importer pay up front, they become aware of the cost of not recycling their product and its packaging. The manufacturer and importer have the ability to change the product and thereby reduce the tax they must pay and pass on as increased cost of their item. The consequence of an action needs to be with the entity responsible for the action and can thereby change the action to avoid the consequence. Punishing someone who has no responsibility for the original action only penalizes someone with no chance to change the action.

The problem with recycling is that some of it is actually expensive and, in some cases, unrealistic. These issues impact the environment and will become bigger issues in the future. So how do we deal with them?

To encourage recycling, local municipalities should not weigh the recycling, but require the recyclers making the pickup to confirm that recycling meets all requirements. If not, it won't get picked it up. If people put garbage in the recycling bin to avoid paying the garbage fee, reject the whole bin. If the recycling is dirty or contaminated, reject it. Put the burden on the consumer to follow the rules. They don't have to pay the garbage fee on recycled material, but it must meet the recycling requirements. Allowing consumers to throw anything into the recycling bin without following the rules means that a worker is required to sort the recycling at the recycling center, which makes recycling more expensive and less likely to work.

Further, composting and food recycling should be encouraged. There are already plenty of community and home composting options available. But if someone puts out composting that is contaminated with other material, reject it. In a few areas, local farmers are collecting food scraps to feed livestock, another program that should be encouraged. The whole idea is to make our society more ecologically friendly in as many ways as possible without requiring massive governmental intervention. And that means putting the requirements and efforts on individual citizens and consumers.

Please note: we aren't preventing people from creating masses of garbage and continuing to be wasteful; we are just making the consequences for that action much more immediate.

Section VII: Political Policies

Political (adjective): 1. relating to the government or the public affairs of a country, a period of political and economic stability. 2. relating to the ideas or strategies of a particular party or group in politics, a decision taken for purely political reasons. 3. interested in or active in politics, "I'm not very political." 4. motivated or caused by a person's beliefs or actions concerning politics, "a political crime."

—Oxford Languages

September 18, 1787: A lady asked Dr. Franklin, "Well, Doctor, what have we got, a republic or a monarchy?"
"A republic, if you can keep it," replied the doctor.

—James McHenry

37.
Gerrymandering

Political gerrymandering makes the incentive for most members of Congress to play to the extremes of their base rather than to the center.

—Barack Obama

The word *gerrymandering* came into being thanks to the *Boston Gazette* on March 26, 1812, and was written as *Gerry-mander* to describe a redrawing of state senate districts authorized by Governor Elbridge Gerry. There was one that looked like a salamander that favored the Democratic-Republicans who were trying to increase their hold on the Massachusetts legislature. There is no proof that Governor Gerry helped shape the odd-looking district, but newspapers aligned with the Federalist party were quick to latch onto the new word and republish it enough times that it gained an entry in the Oxford dictionary: "Gerrymander (verb): manipulate the boundaries of (an electoral constituency) so as to favor one party or class. Achieve (a result) by manipulating the boundaries of an electoral constituency."

With computers and the aid of detailed voting and census data, politicians and their analysts have taken gerrymandering from an old political art into a science of political manipulation. Some modern districts make the original salamander shape seem very tame by comparison. Legal challenges to some of the more outrageous examples suffered a major setback with the June 2019 Supreme Court ruling that the question of partisan gerrymandering was a political one and not a legal one, best resolved by the elected branches of government. The problem with referring to it as a political solution is that any party controlling a state legislature has every reason to slant the districts in their favor. With a distinct advantage to be gained by a party redrawing the districts with a slight edge and packing all their opponents into a few districts, it's not going to change. This process is called "packing and cracking," where the opponents are "packed" into a few voting districts and all the surrounding districts are "cracked" to give the controlling party an edge. This isn't just a Republican issue; the Democrats have done it as well. The Republican party made a concentrated effort before the 2010 elections to gain control of state legislatures prior to the 2010 US census data changing the apportionment of seats in the House of Representatives.

In Wisconsin, the 2010 redrawing controlled by Republicans resulted in a super-majority within the legislature. After the 2020 election, Republicans held sixty-one of the ninety-nine seats in the legislature. Looking at only the contested seats where a Republican ran against a Democrat (twenty were uncontested—thirteen Republican, seven Democrat), the data from the Wisconsin Secretary of State shows that Republicans won forty-eight of the seventy-nine contested seats (77%) with only 51% of the combined votes in the contested races. This shows that Democrats are more packed into contested districts while Republicans have been cracked into enough districts to give them 77% of the seats. The Wisconsin State Senate was even worse, with only sixteen seats in play out of thirty-three. The Republicans won ten (63%) of the sixteen with fewer total votes than the Democratic candidates, giving them a twenty-seat Republican majority.

The Wisconsin election in 2018 showed an even starker difference, where Democrats won all statewide offices, and 55% of the votes cast in legislative assembly elections, but only won 36% of the seats in the assembly. For the assembly elections, twenty-nine Democratic seats were uncontested, meaning that the districts were so favorable to Democrats that Republicans chose not to even run. Only four Republicans ran unchallenged for assembly seats in 2018. Republicans won 89% of the contested seats with only 59% of the votes in the contested districts.

In most states, the legislatures also draw up the House of Representative districts. For Wisconsin in the 2018 House of Representative elections, the Republicans got 46.8% of the votes, but with gerrymandering, the Republicans won five of the eight House seats. The reason to focus on Wisconsin is that they were part of the 2019 ruling on gerrymandering by the Supreme Court. Also, the Wisconsin Secretary of State supplies election results in Excel spreadsheets, making analysis of the data very easy. Having said that, there are Democratic states that have gerrymandered districts as well.

This is not a liberal versus conservative issue; it is about fundamental fairness and respect for a changing society. If the political party in control is using gerrymandering to retain political control even as a region shifts toward the other party, then this is fundamentally wrong. It creates an oppression and lacks respect for most of the voters, even as the party in control proclaims its respect for democracy. The core problem is when a political party tries to tenuously hold onto power without recognizing that society has changed, and their beliefs and policies are no longer in step with the majority.

When a party in power finds that society has changed, and the party hasn't, its leaders need to step back and take a fundamental look at their beliefs and policies. To use gerrymandering to hold power is a belief in the win-at-any-cost model, but it doesn't recognize when society has changed. And a win-at-any-cost model can rapidly devolve into authoritarianism. This isn't the first time society has changed and left an American political party to decide to either change or fade away.

It should be noted that the Supreme Court's majority verdict written by Chief Justice John G. Roberts Jr. and joined by four other conservative justices, said that in order for judges to evaluate claims of partisan gerrymandering, they would need "a limited and precise standard" that would be "clear, manageable, and politically neutral." But no one had proposed one, the court said.

The solution is to find that neutral standard and hopefully a nice little automated computer program that allows us to draw voting districts according to something that is non-partisan with basic rules. One of the measures of how gerrymandered a district is how often zip codes are broken across a voting district's boundary. The more partial zip codes included in a district, the more gerrymandered it is; this is a yardstick used by political scientists to measure the degree of gerrymandering. One solution being discussed among political scientists is to require that zip codes not be split and must be included in a district whole. One paper discussed this idea, identifying that while 82% of zip codes are completely in a single congressional district, the remaining 18% are sometimes split between two to five.[104] One map showed three zip codes, each one split across four congressional districts and showed that approximately 22.5 million Americans live in areas where they are in a different congressional district than the majority of the people in their zip code.[105] This may not seem like much, but correcting this could be a first step in reining in the gerrymandering that is creating a flaw in the basic American view of majority rule.

The proposed policy is that a voting district has to fully encompass a zip code, except where the number of registered voters in a zip code is more than half the size of the required number of registered voters in the voting district. Usually, the size of the district is determined by the number of registered voters in an area (state, city, etc.) and dividing it by the number of voting districts to determine the target number of voters per district. The problem is that when drawing up small voting districts for local elections, there may be zip codes in which the number of voters is too large to allow the policy to be maintained.

The policy would also include a provision that zip codes cannot be redrawn to reflect the voting habits of the constituents. That may seem obvious, but unfortunately failing to stipulate what might seem obvious has allowed people to bend the rules and game the system. A policy should always be written with the thought of how people could misuse it.

38.
Voting Laws

The future of this republic is in the hands of the American voter.

—Dwight D. Eisenhower

Voting rights is a major issue, with Democrats claiming voter suppression and Republicans claiming voter fraud, or a need for more election security.

The first thing we need to understand is that the charge of voter suppression goes back to after the Civil War, the era referred to as Reconstruction when the states that had been the Confederacy were undergoing massive changes. Many people were pushing for enfranchisement of Black voters, and the existing Southern political structure was fighting back to retain their power. The result was the creation of the Ku Klux Klan as a force of terror and voter suppression laws that required Black voters to attempt and fail tests that many white voters would never have passed. This was the beginning of the Jim Crow era where racism was rampant. The suppressive laws lasted well into the 1960s, when efforts like the Freedom Riders and Voter Registration drives confronted the laws and practices designed to keep Black voters disenfranchised. The present-day charge of voter suppression is focused on the lack of access or actually voting, to registration voter ID laws, and purges of voting rolls.

The issue of access for voter registration is a lack of places to register to vote in low-income areas. But if grassroots organizers are able to mount voter registration drives, then these charges are difficult to sustain. The usual charge is about closing Department of Motor Vehicles offices in minority neighborhoods, but with the local organizers able to register voters, that seems ridiculous and shouldn't be used to limit a state government's decision of where DMV offices are located or if they are reducing the number of DMV offices due to state budgets or lack of use.

Another issue is a request that voters can register at the poll and vote that day. This invites the potential for fraud charges from one side and lack of registration access from the other. The question is how far a voter should have to go to register and how much documentation is required to prove the voter lives in the district and is a citizen.

But this also involves the potential voters' failure to register, a failure that lies with the voter and not the government. Government must provide the ability for a voter to register, but if the voter fails to exercise that ahead of an election, whose fault is that?

So, what is the minimum that government needs to provide to allow people to register?

First, there is a practical requirement that the voter has to be identified as a qualified citizen and where they reside to determine what elections they can vote in. With a state-issued identification card (driver's license, state ID, etc.), that is fairly easy if you trust the address on the ID. The problem is that in this hectic economy, people move, and some lack a state ID. To make it even more confusing, there is no federal database identifying who is a US citizen. The Social Security Administration has a comprehensive database, but no one is required to get a social security card. The State Department knows all the passports and visas they have issued, but again, it is not required for citizens to have passports. Some may see this lack of a central database as a problem, but this is one of those issues where the responsibility lies with the states, and a federal database could be seen as overreach.

The requirements for verifying whether someone is a citizen and where they live varies from state to state, and there is a non-profit organization, the Electronic Registration Information Center (ericstates.org), which shares voter registration data among states to allow them to catch people who move or try to vote in different states, and to increase the accuracy of the voter rolls. But ERIC is voluntary, and currently only thirty-one states and the District of Columbia participate. The problem is that after Donald Trump publicly attacked it, some Republican-led states are now withdrawing from it. The key issue is that ERIC produces reports on people who are eligible to vote and requires states to contact them, something that some Republican officials have called superfluous and a waste of resources, hence their move to withdraw from ERIC.

The National Voter Registration Act of 1993 and the Help America Vote Act of 2002 required states to allow voters to register when they applied for a driver's license or public assistance, or when they mailed in a registration form. There have been various challenges to the laws, but so far, the Supreme Court has upheld them.

So, what needs to be done? Just continue to monitor and make sure no state passes something that conflicts with federal voting laws. While ERIC could be a major help to election security, mandating participation is not the answer, and the present stance of some states reflects the political winds that are blowing right now.

Second issue: what is the responsibility of the voter? This is the big question because the issues above have mainly sprung from potential voters not registering before an election or voters allowing their registration to lapse. It comes back to this: if voting is important to you, then you should put in an effort to ensure you are registered prior to the election by whatever time frame the state sets.

The second charge is lack of access to voting. This is easily provable if voting involves long lines and delays that working people cannot tolerate. Those leveling the charge need to take the data for each district and compare where voting locations are in relation to the district and how many machines are available at each location. If the distance to the voting location or the number of voters per machine varies in a pattern that falls along the party lines of the majority of the voters in the district or along racial lines, then the charge could be valid. If Republican districts in a region require less travel and have more voting machines than Democratic districts in the same area (state, congressional district, etc.), then that is voter suppression. But as of the year 2023, no such charge is supported. That doesn't mean it won't be presented. But it does mean that activists need to stop leveling these charges until they show the analysis that proves that voters are being slighted in their ability to vote. Now, the flip side of this is that every state Secretary of State should already have this analysis and know if there is a discrepancy within voting districts and work to correct it. If some districts have fewer voting machines per voter than others, that Secretary of State should be aware of it, even if it is the responsibility of county election officials. A proactive move by the state agency that administers elections is far better and cheaper than fighting activists in court and finding out the hard way they were right.

The newest front in this battle is the argument to withdraw voting locations from college campuses. While some Republican activists are making this argument, it goes back to the issue where younger voters vote predominantly Democrat. Removing polling places from college campuses should provoke lawsuits, especially if the activists can show through basic analysis how many voters were affected with longer distances and longer wait times.

Voter ID laws have been a major push from Republicans to prevent voter fraud. This is contradicted by the fact that there has been little evidence of voter fraud, and this is something most experts don't consider to be an issue. But that has not stopped some Republicans from claiming voter fraud and saying we need stricter voter laws. Those laws tend to focus on what constitutes a valid ID at the polling locations to allow a person to vote. The concern from Democrats is people who no longer drive, have never gotten a passport, or don't have a state ID—how do they vote? In Texas, a voter ID law requiring a driver's license, passport, military identification, or gun permit, was repeatedly found to be intentionally discriminatory by courts. While the Department of Justice under the Trump administration had viewed the Texas laws as reasonable, there should be a way for people who are able to vote, but don't have these forms of identification. But there is also a reality that people who come to vote need a way to positively identify themselves to poll workers. While voter fraud has been incredibly rare, it has occurred. So, what is the solution?

Like so many other cases, there isn't a one-size-fits-all answer, and while voting is a state responsibility and each state (including DC and Puerto Rico) will need to enact their own laws, there should be some federal guidelines. So, if you have a state-issued ID with a picture on it, the poll worker confirms you are on the voting list and checks the photo ID, which allows you to vote. If you don't have state-issued photo ID, then there should be a station operated by a poll worker, where the person lacking an ID can sign, have their picture taken, and then vote, provided they are registered. The signature is kept with the photo, and the next time they vote, the poll workers can just use the previous photo. Facial recognition has gotten to the point where a tablet computer can easily confirm that someone is the same person. When a person registers to vote, their picture can be taken to start the recognition process, or it can be done the first time a person has voted. And it has an added benefit. Forgot your wallet? No problem, step right over and smile for the camera.

Think this is government overreach, possible gathering of facial data in preparation for a totalitarian takeover? Then bring a state-issued photo ID. The idea of a government having your photo shouldn't bother you; it allows the government to know who a citizen is. That is a basic requirement of a democracy, and with the present system, adding a photo only makes it easier to confirm who is allowed to vote and ensuring that elections are fair.

Purging voter rolls is something that almost all jurisdictions do, usually based on records of who has died or moved. In nine states, if you haven't used your vote in recent elections, it is removed—the so-called "use it or lose it" states. The problem is in some locations, the attempt to notify the voter fails, and worse, the voter who has suddenly returned to vote is not allowed to re-register as the last day to register to vote has passed. The other problem is when purges are done with questionable reasons. Two examples happened in recent years. In July 2017, with a Georgia gubernatorial election about to take place and the Secretary of State Brian Kemp in the running, the Secretary of State's office removed half a million Georgian voters from the voting rolls. Democratic challenger Stacey Abrams declared foul, and the controversy was on.

The other example is when Wisconsin wanted to hold off removing 234,000 voters before the 2020 election to give them the opportunity to vote because they had found that the change of address records from the DMV were unreliable. Conservative activists sued the state, demanding that the 234,000 be removed immediately.

The core issue here is the voter's personal responsibility. If a voter does not vote for multiple elections, do they have a right to walk into their polling place and expect to vote? The judicial rulings have been inconsistent on this subject, with some local judges supporting purges and some rejecting them. The US Supreme Court voted five to four in 2018 to support Ohio's "use it or lose it" voting law. Ohio sent notifications to voters who were about to be purged to confirm they still lived at the same address. Plaintiffs in the case indicated that they had not received notifications and weren't allowed to vote because the registration deadline had passed. This contrasts with a number of states where automatic voter registration (AVR) occurs. Oregon passed the first AVR law in 2016 and about fifteen others adopted similar measures. The solution might be the same as the Voter ID. If the record still exists and the signature and photo match it, let the person vote. It isn't a registration; it is a reinstatement of the existing record.

The question of voter purges leads into the next item—the recent claims of voter fraud. There have been many claims of voter fraud, but little evidence of it, and in some cases, the evidence is hearsay or even fabricated. The Republican argument that had some validity was that officials in 2020 had changed voting laws just before or as mail-in and early voting had begun. That charge ignores that election officials were reacting to issues during primary elections. Primary elections had been held during the pandemic, which had fundamentally altered the conditions under which elections were taking place. Changing rules in the face of a crisis to try and help elections proceed is something officials should be allowed to do.

Unfortunately, the challenges couched these changes as trying to sway the election for Biden even when they were done by Republican officials. But it does highlight an issue: the lack of consistency between states. How different states conduct elections is something that stands out in this country. Each state has different voting laws and different methods on how voting is conducted.

Oregon passed Ballot Measure 60 in 1998, which made vote-by-mail the standard mechanism for the state. Since then, Oregon consistently ranks as a national leader in voter turnout. That contrasts with several states that had issues expanding mail-in ballots. Five states require that counting mail-in ballots can only begin on election day. Three of these, Pennsylvania, Michigan, and Wisconsin, were battleground states in 2020, which fueled some of the unfounded calls of fraud. Being unable to count mail-in ballots until election day resulted in polling results slowly changing on election night and over the following days as mail-in ballots were counted. If prompt results are required, then those states need to count the mail-in ballots prior to election day as many other states already do.

The differences between states leads to confusion. How an election is conducted is entirely within each state's purview. The only suggestion is to convene a meeting of all the secretaries of state and generate a set of national guidelines, especially on mail-in balloting. These guidelines would be purely voluntary. States like Oregon could show how they conduct elections and make mail-in balloting safe and effective, and the other states could learn.

One issue in the present round of voting laws is changing how elections are certified and the procedures for nullifying and overriding an election. This is especially difficult as the 2020 elections were declared safe and secure by all the people who actually ran the elections, Republicans and Democrats. But the public insistence that the election had been stolen led people to believe that and call for changes. A lot of these are minor issues or nuisances, but enabling legislatures and partisan committees to rule on the validity of an election has opened the door to future abuse. We need to lay out a set of guidelines on how and when an election can be declared invalid. This should be generated by a meeting of the secretaries of state with help from their election officials to develop a federal set of requirements.

39.
Dark Money and Politics

If you're a lobbyist who never gave us money, I didn't talk to you. If you're a lobbyist who gave us money, I might talk to you.

—Mick Mulvaney

The Citizens United Supreme Court decision in January 2010 prohibited the restriction of independent expenditures for political campaigns by corporations, including non-profit corporations, labor unions, and other associations. "Independent expenditures" in this context means a political campaign that expressly advocates for the election or the defeat of a clearly identified candidate. This expenditure is not made in cooperation with or at the request of a candidate or their committee or political party. This opened the door for political action groups to weigh in on campaigns for or against candidates provided they were not coordinating with any candidate in the race.

Prior to this ruling, these organizations were prohibited from making any "independent expenditures" thirty days prior to a primary election or sixty days prior to a general election. Not that they couldn't at all, but they couldn't in those crucial days right before the election. And the Supreme Court ruled that based on the free speech clause of the First Amendment, the laws blocking those expenditures before an election were unconstitutional. The vote was five to four, with all the justices appointed by Republican presidents voting to declare the laws unconstitutional, and thus began the days of the political action committees (PACs), Super PACs, and dark money.

Dark money refers to the ability of political action groups to take in cash and then spend it on political activity (advertising, flyers, etc.) without reporting where the money came from. The laws were changed in 2018 so that any PAC that makes "independent expenditures" must now report its donors, and are limited to spend 50% or more of their expenditures on non-political work.

So, how do you bring in money, keep donors anonymous, and still spend the bulk of the money on politics? The solution is a web of PACs. The top tier takes in money from donors who wish to remain anonymous. This top tier doesn't do any political work; they don't report their donors. They provide the money as grants to other PACs. This second tier of PACs makes grants to other PACs on the second tier. As the money gets swapped around, it becomes a bigger portion of the operating money of the PACs and lets them spend more on political work.

Let's say the top-tier PAC gives $5 million to PAC A, which then gives $4 million to PAC B. PAC B spends $1.5 million on political work, which is less than 50% of the $4 million and then passes the remaining $2.5 million back to PAC A via an intermediary, PAC C. PAC A then spends $3.5 million on political work, which is less than 50% of the $7.5 million that has flowed through its doors. All $5 million gets spent on politics and none of the donors to the original top-tier PAC have to be identified. And by the way, those top-tier donors can't be foreign entities, which is totally illegal since there are laws against foreign interference in US elections, right?

Allow me to introduce the case of Lev Parnas. Anna Massoglia documents how Parnas, a Soviet-born former business associate of Rudy Giuliani, funneled foreign donations through straw donors, shell companies, and multiple PACs to get the money into US elections. [106] And that is one of many examples that have come to light, worked their way through the US legal system, and ended up in convictions because the involvement of foreign funds made it illegal. If Parnas or any of the others had used funds from US citizens, US labor unions, or US corporations, it would have been perfectly legal.

The poster story for legal dark money was the recent gift of $1.6 billion to the Marble Freedom Trust. While the donor's name has been revealed via journalists, a complex arrangement still allowed Barre Seid to transfer his company to the Marble Freedom Trust, sell it to an Irish multinational for $1.6 billion, and avoid paying an estimated $400 million in taxes. Marble Freedom Trust now has the entire $1.6 billion to spend on conservative causes, and the IRS got nothing on the deal.

But don't think this is limited to conservatives. Democrats may have been slow to embrace dark money, but they stepped up in 2020 and raised and spent something on the order of $1.5 billion in dark money.

The first federal campaign finance law was in 1867, and it prohibited federal officers from requesting campaign contributions from Navy Yard workers. What followed was a hundred years of tightening campaign finance laws with some of the high points being:

- 1907: After a corporate fundraising scandal in his own campaign, Teddy Roosevelt signed a law that banned corporations and national banks from contributing to campaigns.
- 1910: The Publicity Act set limits on contributions to candidates and required public disclosure in writing of spending by political parties and candidates in federal elections.
- 1935: Congress banned public utility companies from making contributions.
- 1990: The Supreme Court upheld that for-profit corporations and non-profits like the Michigan Chamber of Commerce (plaintiff in the case) were still banned from campaign contributions.

There is a lot more detail to the hundred and thirty years, but the big takeaway is that the years between 1867 and 1990 limited corporations and wealthy individuals' ability to affect elections. The Citizens United decision in 2010 overturned most of that in one day. In 2014, Shaun McCutcheon and the Republican National Committee challenged the legal biennial limit ($123,000 in 2013) on the amount an individual could contribute to all candidates, PACs and parties combined. The same five

Supreme Court justices who ruled in Citizens United ruled that the limit did little to stop corruption, while significantly restricting participation in the democratic process. With the limits gone, every election since has outspent the previous one, and the influence of corporations, PACs, and mega-donors has exploded. In the 2008 election cycle, all the candidates for Congress raised $2.1 billion that was spent on congressional races. In 2022, it was $5.2 billion.

Why did Supreme Court justices appointed by Republican presidents feel the need to give corporations and wealthy individuals the ability to donate massive amounts of money to campaigns? The congressional winners of the 2008 cycle raised $1.6 billion of that $2.1 billion, so money translates to victory. In the 2022 cycle, the sitting members of Congress raised $3.7 billion of the overall $5.2 billion. Wonder why people say that congressmen spend half their time on the phone with donors? It's because the data shows that the person with the biggest war chest wins. So, if you are losing, you need to figure out how to get more people to give or allow rich donors to give more. The Supreme Court, in their 2010 and 2014 decisions opened up the ability of PACs to influence elections right to the end and the rich to give more, claiming it was their First Amendment right to outspend everyone else.

Again, this is not just a Republican problem. In the 2020 cycle, Democrats raised and spent more dark money than Republicans.[107] This is an American problem. We have opened the door to wealthy corporations and individuals having a proportionally larger voice in American politics. Historian Heather Cox Richardson chronicled the formation and rise of the Republican party and found that the Republican party had cycled between two main ideologies. Founded in 1854 as a way to prevent new states entering the union from becoming slave states, it sought to provide an equality of opportunity for middle class farmers to develop in these regions instead of being dominated by large slave-powered plantations controlled by a few wealthy individuals. This idea of equal access of economic opportunity was enshrined within the Declaration of Independence, she argues.[108]

But after the Civil War, the party morphed and focused on the protection of property rights, a concept defined in the other great document of American government, the Constitution. When a party is dominated by the wealthy, they focus on protecting their property rights and maintaining their wealth. When you are on top, your focus is on staying there. A progressive income tax, higher taxes on corporate earnings and programs that remove wealth from the top 1% and disperse it to the bottom 25% all run counter to the principle of property rights. The present Republican party, for all its talk of Trump populism, is firmly in the protection-of-property-rights camp.

And we return to a central point of this book: that two things can be opposite and true at the same time. America needs both protection of property rights as enshrined in the Constitution and equality of opportunity as defined in the Declaration of Independence.

And that is a fundamental problem with allowing corporations and labor unions to spend unlimited money on politics. They don't vote; they can't run for office. They are legal entities that exist only on paper but can own and control physical assets through the people who run them. In 1905, President Theodore Roosevelt said during his address to Congress, "All contributions by corporations to any political committee or for any political purpose should be forbidden by law." This was in the days of the "robber barons," when corporations had grown large and controlling and were seen as crushing the little guy. Roosevelt is remembered for breaking up trusts and monopolies that allowed a few men to control massive parts of America's industries. One has to wonder what he would think of modern America.

The Tillman Act was passed in 1907. Its language was simple—it prohibited national banks and federally chartered corporations from contributions to election campaigns at any level and went further in prohibiting "any corporations whatever" from spending money on presidential or congressional elections. Acts in 1943 and 1947 extended this restriction to unions.

The 2010 Citizens United decision overturned all of this. In his dissent, Justice Stevens wrote:

> In the context of election to public office, the distinction between corporate and human speakers is significant. Although they make enormous contributions to our society, corporations are not actually members of it. They cannot vote or run for office. Because they may be managed and controlled by nonresidents, their interests may conflict in fundamental respects with the interests of eligible voters. The financial resources, legal structure, and instrumental orientation of corporations raise legitimate concerns about their role in the electoral process.

This concept can be taken a step further when the wealthy are added to the mix. By allowing those with more property rights (the wealthy) to spend an unlimited amount on politics, we have fundamentally shifted the balance away from equal opportunity toward the protection of property rights. To then provide those same wealthy individuals and groups with the ability to hide their donations through a shifting network of PACs and SuperPACs, we have cloaked them in anonymity and protected them from any consequences.

A CBS News poll in 2022 seemed to confirm this.[109] In the poll, 72% of respondents indicated that they felt democracy was threatened, and 86% said that the influence of money in politics was the major threat. Democrats have now taken up the cause, but with Republicans digging in their heels, where will this go? And what can be done?

The problem is that the current Supreme Court has declared that restrictions on political donations by the wealthy, corporations, and unions cannot be restricted except to political campaigns (Tillman Act of 1907). Justice Clarence Thomas went even further, arguing that any restrictions were unconstitutional. So, the problem starts with the Supreme Court. As long as the majority on the court favors the position that more wealth allows more political voice via spending, then there is very little that can be done. After the 2010 decision, Senator John McCain, who authored multiple bills on campaign finance reform, stated that "campaign finance reform is dead." He also predicted a backlash against the massive flow of money. In an appearance on *Face the Nation*, he noted that the justices were far removed from the realities of campaign fundraising. He said:

> I think that it was interesting that they have had no experience in the political arena. I was reminded of the story of Lyndon Johnson, when he was vice president, was told about President Kennedy's appointments of all these brilliant people, and he said, "You know, I wish one of them had run for county sheriff."

While some have proposed expanding the Supreme Court, that would be a short-term solution that could create long-term headaches. The reality is that the efforts by conservative legal groups have packed the court with conservative jurists who have struck down campaign finance reform, as well as *Roe v. Wade*. They are strong believers in property rights, and until there is a retirement or death, this is how it will be.

Others have proposed a constitutional amendment, and that is the path that needs to be taken. Wealth and corporations have an overwhelming ability to influence politics, which was recognized and acted on in the 1900s. But those actions were not enshrined in the Constitution, and as a result, we are back to square one with the wealthy able to unduly affect the political landscape. The last amendment was ratified in 1992. It's been over thirty years since anything has changed in the Constitution.

This won't be easy. A proposed amendment must be sent to state legislatures for approval through Congress by a two-thirds super-majority or via a national convention called for by state legislatures. Congress, with an almost fifty-fifty split, is a difficult route, and a national convention called by state legislatures has never happened. Further, once in the hands of state legislatures, thirty-eight must approve to take effect, and that means getting Republican- and Democrat-controlled states to vote for it. Difficult, but this is the way to resolve the issue with the present make-up of the Supreme Court.

So, if you can't stop the flow of dark money, what do you do? The answer is to increase the voters' involvement. Town halls and debates are the antithesis of dark money. This is how voters review and meet candidates without the influence of dark money. The trick is to find an independent third party to sponsor the event and act as the referee, such as the League of Independent Voters, a local civic group, or a local university.

We expect our candidates to walk up to the podium with nothing at a debate, requiring them to memorize facts and data in addition to being ready to discuss the issues. That is a bit of an ask. In this modern world, requiring rote memorization of facts is a skill that politicians only use on the campaign trail. And let's be honest: you are not just electing the candidate; you are electing their team as well. Candidates should be allowed to prepare a data book to help them with all the facts and figures and make things fair. Those books need to be shared ahead of time with the organization acting as the referee as well as with their opponent. Why? We can have different opinions, but we all need to work from the same facts. Sharing the data book with the referee and opponents allows everyone to fact-check ahead of time, so if someone tries to say the earth is flat, their opponents and the moderators are ready to call them out on the misleading data.

Think that's extreme? As I write this, a GOP state chairman has come out in favor of the theory that the earth is flat. Which seems very weird when Greek philosopher and mathematician Eratosthenes published not only that the earth was round but estimated its circumference in his book *On the Measure of the Earth* in 240 BC. How accurate his estimate of 250,000 stadia was, is based purely on which length you use for the ancient measurement of a stadia, which historians estimate was anywhere between

150 and 210 meters. The actual circumference is 40,075 kilometers, which means if the stadia was a little over one hundred sixty meters, Eratosthenes would have been accurate. Considering he was using sticks, the angle of shadows, and hand calculations over 2,200 years ago, it is still pretty amazing that he got an answer that was very close to the actual circumference.

The point here is that all candidates should walk into a debate with a single set of facts, and it becomes a debate about the issues and how each candidate would approach them. But this also means that voters need to get more involved and take an interest in the process to help counter the effects of dark money.

As a final point to money in politics, the last amendment to be ratified is the twenty-seventh: "No law, varying the compensation for the services of the Senators and Representatives, shall take effect, until an election of representatives shall have intervened." Seems totally reasonable that Congress can't enact a pay raise until voters get to give Congress a review via elections. It was originally proposed in 1789, introduced by James Madison as one of several amendments in the very first session of Congress. It had languished for over two hundred years and was ratified by only nine states until it was stumbled upon by a University of Texas student in 1982. Gregory Watson's term paper pointed out that the amendment had been written without an expiration date and that Wyoming had just ratified it in 1977 as the ninth state, still short of the required thirty-eight states. He argued that it could still be added to the Constitution if more states ratified it. He got a *C* on the paper as his professor questioned the argument. Launching a one-man campaign to bring the amendment back from its zombie status, he spent thousands of his own dollars on a letter-writing campaign over the next ten years that resulted in Michigan becoming the thirty-eighth state to ratify it on May 7, 1992.

40.
The Fifty-first Star

I think Puerto Rico becoming a state would fulfill the destiny of 3.5 million American citizens that live in Puerto Rico.

—Ricardo Rossello

There has been a lot of talk about making Washington DC the fifty-first state. But it is really a federal city with a land area of 61.05 square miles, a population of almost 700,000, and the ability to vote for president. Changing it to a state only enables it to have two Senators and a representative in the House. As the city voted for Joe Biden by 89% in 2020, it is clearly a move by the Democratic party to add to its position in Congress. The objection to this is that the District of Columbia was carved out of Maryland as a federal district to be administered by Congress as defined in the Constitution. It used to contain land in Virginia, but this was returned in 1846 when the city of Arlington petitioned to return to Virginia due to charges of mismanagement from Congress over the area on the southern bank of the Potomac. There have been multiple proposals over the years to either make DC the fifty-first state, have everything but the land under the Capitol and federal buildings returned to Maryland, or to simply allow the residents to vote for congressional representatives in Maryland. The Democratic push for a fifty-first state might seem reasonable, but the DC area has local representation (the mayor and city council) and can vote in the presidential election. Giving them two Senators and a House representative would only provide an advantage for the Democratic party. If its lack of representation in Congress is the issue, then let them vote in Maryland. Adding Washington DC's population to Maryland would end up with an additional representative for Maryland, and if it was stipulated this new House district could be within DC, that is a compromise. But the federal district is established in the Article one, section eight of the Constitution, and the Twenty-third Amendment established the electoral votes for Washington DC, so I really think this isn't a vote in Congress, but a constitutional amendment that is required to either make Washington DC the fifty-first state or have them vote in Maryland for Maryland's senators and add an additional House seat.

I think requesting a decision from the Supreme Court would be wise if a constitutional amendment is required to allow a congressional representative district to be established in Washington DC as part of Maryland and allow DC voters to vote in Maryland federal elections. And if the court so rules, then start the process.

Puerto Rico, on the other hand, is a different story. It has a population of about 3.2 million, a larger population than Iowa. And it cannot vote for president and has no representation in Congress. A non-voting representative is presently allowed to sit in the House of Representatives much like Washington DC's. Puerto Ricans are citizens of the United States and can move freely between the mainland and Puerto Rico, but the difference is they cannot vote in federal elections in Puerto Rico but can by moving to the mainland.

The United States acquired Puerto Rico in the aftermath of the Spanish-American War of 1898. Spain ceded Guam, Philippines, and Puerto Rico as part of the peace treaty to end the war. The United States ruled that Puerto Rico was a quasi-colony with a governor and upper house of the legislature appointed by the United States, but the lower house locally elected. The lower house unanimously voted in 1914 for independence. This was rejected, but it created an opposition to US rule, which continued to exist even though the United States granted more local control in 1947 and Congress authorized a local constitutional convention to allow Puerto Rico to determine if it would become a commonwealth. In 1950, the opposition went so far as to attempt a revolt against the United States starting on October 31, 1950. This included an attempted assassination of President Truman on November 1 in Washington DC. It was put down by US forces and then in 1952, Puerto Rico voted to become a commonwealth.

In July 2009, the United Nations Special Committee of Decolonization approved a resolution calling on the United States to allow Puerto Rico the right of self-determination, something that many Puerto Ricans have been calling for. The election was held in 2012, and in two separate elections, the first asked "should Puerto Rico continue in its present relationship of commonwealth with the US?" The result was no by 54% to 46% to remain as is. The second election asked what form the new relationship should take—statehood, sovereign free-associated state, or independent state. The vote was 61.16% for statehood. Nothing was done as a result of these elections.

On November 3, 2020, the US presidential election was not on the ballot in Puerto Rico, but a simple up and down vote for Puerto Rican statehood had been added by Puerto Rican Governor Garced. Of Puerto Rican voters who participated in the election, 52.5% voted for statehood. Prior to the election, the Justice Department informed Puerto Rico's Commission on Elections that it would not be approving the referendum, meaning all funding would be handled by the Puerto Rican government. One reason for the rejection was that there was no option to vote for continuing to be a territory. Republican leader Mitch McConnell in 2019 said he would not allow the issue to be voted on or discussed in the senate and called it a "government overreach." Many other Republican officials feared that adding Puerto Rico as the fifty-first state would provide the Democrats with another state in their column. A state larger than Iowa would have two senators, four representatives, and six electoral votes.

The problem is that once a territory holds an election and votes to become a state, how can we deny it representation at the federal level? Hawaii voted for statehood and became the fiftieth state in 1959. Hawaii's population is less than half that of Puerto Rico.

This is a problem. The United States is holding Puerto Rico in the same status as Britain held the original colonies in 1776, subject to but without representation in the federal government. The United States either needs to vote in Congress to admit Puerto Rico as the fifty-first state or cut it loose as an independent country. Waiting will only kick the can down the road until it blossoms into a full-blown crisis. It would be better to bring it to the floor of Congress and resolve it to place one more potential crisis behind us.

Yes, changing Old Glory to fifty-one stars will be difficult and painful for many, but far less so than the possible consequences of keeping over 3 million people disenfranchised. The point is we can deal with this now when the situation is calm in a rational and legal manner, or wait until someone in Puerto Rico with possible foreign influence decides it's time to start a revolution. Proclaiming the reasons that necessitated the revolution of 1776 as core American values, but then committing the same sins against Puerto Rico seems the height of hypocrisy.

41.
Electoral College

After immersing myself in the mysteries of the electoral college for a novel I wrote in the '90s, I came away believing that the case for scrapping it is less obvious than I originally thought.

—Jeff Greenfield

The electoral college has become a major issue among Democrats because of the defeats of Al Gore in 2000 and Hillary Clinton in 2016, in which the Democratic candidate won the popular vote but lost in the electoral college. This also happened in the 1824, 1876, and 1888 elections, not to mention a number of elections with three or more candidates. One example was Lincoln's 1860 victory where he won 180 electoral votes of a total of 303, with only 39.8% of the vote, the largest share of the four major candidates.

Many have indicated that the electoral college unfairly favors small Republican states. Some even claim it was included in the Constitution as a way for the Southern slave-holding states to skew the election process in their favor. There is no question that the Three-fifths Compromise did skew it toward slave-holding states. But originally the electoral process had been proposed as an alternative to the original proposal of Congress electing the president, much in the way prime ministers are elected in Great Britain and other democracies. Popular vote was considered, but with the potential for a multitude of candidates and the fastest means of communication being a man on horseback, the founders decided that electors would be selected from local districts and then travel to each state house to select a president. These electors could not hold political office and would not be beholden to anyone but would vote based on the information and their conscience. At least that was the plan. The problem was the Founders hadn't counted on political parties or delegates pledging to a presidential candidate. The original plan was to elect someone to represent the district, but it devolved into voting for someone committed to a single candidate. This changed over time from electing electors within state districts to a winner-takes-all approach. In most states today, win the state and that state commits its entire slate of electors to a single candidate. There are two states, Nebraska and Maine, that still use the district method and allocate individual electors to candidates. The complaint (mainly by Democrats) is that when a candidate wins the popular vote but loses the electoral college, something is very wrong, and that favors the Republican party.

But the Democrats need to consider that for every Wyoming, there is a Vermont; for every Montana, there is a Delaware. There are fifty states and the District of Columbia resulting in fifty-one voting states in the electoral college. Each has the number of votes equal to the number of members it has in the Congress. The Twenty-third Amendment was enacted in 1961 and granted the District of Columbia "a number of electors of President and Vice President equal to the whole number of Senators and Representatives in Congress to which the District would be entitled if it were a State, but in no event more than the least populous State," which is essentially three. And looking at the seventeen smallest, which includes Washington DC, they control sixty-five electoral college votes, which is 12.1% of the college, with only 6.6% of the population and 6.9% of the 2020 votes cast. That compares with the seventeen largest states that have 340 electoral votes, which is 63% of the electoral college with 69.5% of the population, and those states had 69.3% of the votes cast. So, it's a 6% delta when you look at popular vote versus electoral college vote between the smallest and largest states.

Looking at the smallest seventeen states in the last two elections, in 2016, Trump won nine of the seventeen states, which worked out to thirty-six of the sixty-five votes. In 2020, Trump won the same nine states, but only thirty-five of the electoral votes, because Nebraska is not a winner-takes-all state, and one of its votes went to Biden. So, the small states are not really breaking heavily to one side or the other. And if you look at the senators from those states, since the 2014 election, the Republicans have controlled 50% to 53% of the thirty-two seats for those states, which confirms overall these seventeen small districts are not strongly leaning one way or the other.

Looking at the seventeen largest states for both elections shows that in 2016, Trump won ten of the seventeen, but that only gave him 185 of the 340 electoral votes, 54.6% of the available votes. In 2020, this is the area where Biden turned the election. Trump won only six of the seventeen states, getting only 122 of the 340 electoral votes. And if you look at the senators, only twelve of the thirty-four available seats are Republican and even in 2016, only 41% of the senators from the large states were Republican.

So where is the Republican strength? It is in the middle seventeen states. In 2016, eleven of the seventeen states went to Trump with 71.6% of the 134 electoral votes. In 2020, Trump won ten of the seventeen, but by losing Wisconsin by only 20,682 votes, he lost ten electoral votes. Trump still got seventy-five of the 134 electoral votes, working out to 56%. And those seventeen middle states are the strength of the Republican senate with twenty-one of the thirty-four senators on the Republican side of the aisle. Six Southern states—Louisiana, Kentucky, Alabama, South Carolina, Arkansas, and Mississippi—form that backbone with all six voting Republican in the last six presidential elections and all twelve of their senators being Republicans.

So, to overturn the electoral college and move to any other system requires an amendment to the Constitution. An amendment has to be proposed by a two-thirds vote of both the House and the Senate. That is the first hurdle. Assuming that somehow two-thirds agree, then the amendment must be ratified by three-fourths of the state legislatures, or three-fourths of conventions called in each state for ratification. The problem is that twenty states have voted Republican in the last six presidential elections, and they hold 30.5% of the electoral college with only 28.2% of the population. Convincing them to give up that 2.3% advantage is highly unlikely.

This brings up the subject of battleground states. There are twenty-three states that were solidly Republican in 2016 and 2020, and there were eighteen states and the District of Columbia that were solidly Democrat in 2016 and 2020. That leaves nine states in the middle, where the average margin of victory in 2016 and 2020 for either major party was less than 5%. Of those, five actually flipped between the parties from 2016 to 2020: Georgia, Arizona, Wisconsin, Pennsylvania, and Michigan.

Trump won red states by almost 7.6 million in 2020 with an average margin of victory compared to Biden of 15.8%, but this was down from 2016. In 2016, Trump won the red states over Hillary Clinton by almost 7.7 million votes, but the margin of victory was 19.2%. In 2020, Biden narrowed the gap in the red states. In the blue states, it was a similar story. The 2016 margin of victory for Clinton was a little over 11 million votes and 21.5%. In 2020, Biden won those same states by 14.5 million votes and a 23.3% margin of victory over Trump.

But the elections were decided in the nine battleground states. In 2016, Trump won seven of them by a total of 594,378 more votes than Hillary Clinton, an average margin of victory of 1.6%. In 2020, Biden won seven of those same nine battleground states by 98,099 votes, an average margin of victory of 0.22%. Of these nine states, Florida and North Carolina voted Republican in both elections, and Nevada and Minnesota voted Democrat in both elections, but not by much. In fact, looking at the difference between the margin of victory in 2016 and 2020 for all states, Biden's 2020 results were an improvement on average of 3.5 percentage points over the 2016 results, even in red states.

It is interesting to look at the overall data for the red, blue, and battleground states. Red states on average are more rural and have weaker economies than blue states. The battleground states are in the middle for average urbanization and strength of state economy.

But the conclusion is that both sides need to accept that the electoral college is the game that is played every four years, and everyone knows the rules.

42.
Gross Domestic Product and Parties

Let me... warn you in the most solemn manner against the baneful effects of the spirit of party.

—George Washington

After gathering the data for the last chapter and doing the analysis on the presidential and the senatorial elections, I decided to add in one more factor. The 2019 gross domestic product (GDP) data broken down by state and then averaged for the 2019 population of each state supplied by the Bureau of Economic Analysis of the Department of Commerce (ww.bea.gov) was added to see if state financial prosperity showed anything. For this analysis, the District of Columbia was removed from the data. While it does have a larger population than Vermont or Wyoming, it gets credit for a lot of federal spending in the GDP analysis, which gives DC a GDP per person twice that of any state. As such, it was tossed out so as not to skew the data toward the Democrats. Why 2019? Because COVID was a massive disruption to just about everything in 2020, and the results were still felt in 2021. As of the wrapping up of this book, the 2022 data was still being gathered and wasn't available.

The results were somewhat surprising. For the states that voted for Biden, the average was $66,441 GDP/person. For the states that voted for Trump, the average was $55,950 GDP/person. This divide grows when only considering the twenty states that have consistently voted Democrat in the last four presidential elections at $68,716 GDP/person. Coincidentally, twenty states have also consistently voted Republican in the last six presidential elections, and their average is $55,663 GDP/person. Remember the six Republican backbone Southern states mentioned in the previous chapter: Louisiana, Kentucky, Alabama, South Carolina, Arkansas, and Mississippi. Their average GDP per person is $46,741. Mississippi is the state with the lowest GDP per person at $38,966, 43% of the richest state, New York, which is firmly in the Democratic column at $91,102 GDP/person.

While the Republican party may claim to be the party of big business, its voting base comes from the states that generate the lowest GDP per person in the union. The eight states with the smallest GDP per person voted Republican in the last six presidential elections, and the six largest all voted Democrat. The seventh and eighth largest GDP per person are Alaska and North Dakota. Remove their oil revenue, and they drop into the lower half, leaving the top nine states all Democrat.

The gist of this is that the states that have been solidly Democrat appear to be the ones that have succeeded and prospered during the years of globalization, and the states solidly Republican have not. But that assumption needed to be checked, so the data for GDP per state and population from 2000 was added and the analysis was rerun. The results proved that the prosperity theory was slightly wrong. The core twenty Republican states have had a 186% increase in GDP per person, where the twenty core Democrat states only increased by 177%. Now, the data showed that North Dakota in 2000 prior to the development of the Bakken Shale oil fields was ranked forty-fourth in the country for GDP per person and has now jumped to seventh, a climb over thirty-seven other states, thanks to that oil revenue. Even removing North Dakota's massive improvement in GDP per person, the Republican states still had improved more.

But then look at the ten states that have changed between the parties in the last twenty years, and the results have a twist. We are going to refer to these as *flip states*. The flip states include the five states that voted for Biden in the last election after voting for Trump in the previous one—Pennsylvania, Georgia, Michigan, Arizona, and Wisconsin; four states that historically voted Republican, but voted for Obama at one point—Indiana, Florida, North Carolina, and Idaho; and finally Iowa, staunchly Democrat but in the last two elections, it has gone Republican and taken their two senate seats with them. But the interesting thing about these ten states is that when ranked by GDP per person, they are in the middle of the pack. The best ranked is Pennsylvania at twentieth of fifty, and the worst ranked is Arizona at forty-first of fifty. But the startling thing is that when you look at the change between 2000 and 2019, the ten flip states have an average improvement of 170% versus the 186% average for Republicans and 177% average for Democrats. Pennsylvania and Iowa have actually improved since 2000, but all the other flip states have lost ground, Michigan being the worst as eighteen other states have surpassed it.

While there are some outliers, the analysis shows that the Democrats' strength are the states with the largest GDP per person, while Republican states are the ones with the worst GDP per person and the flip states that form the battleground are in the middle.

43.
Small-town America's Main Street

Rapid imperial expansion during the middle Republic strained nearly every aspect of the Roman system but none more so than the very foundation of Roman military strength—the small farmer. Spoils of war were channeled into agriculture by the landed elite, resulting in economic polarization and the displacement of independent labor in the countryside.

—Jack Morato

You may be wondering how we got from small-town America's Main Street to the small farmers of ancient Rome, but there is a method to this madness. And it starts with not flying to destinations, not driving only on interstates, and travelling backroads and going through small towns. This has become a habit of mine. I drive the backroads; I stop in small towns. I look around and I talk to people. And the results are rather depressing. Larger towns and a few of the small towns have been able to hang onto their Main Streets. There are some restaurants, hopefully a few stores, maybe even an art gallery, and there are people. But for too many small towns, Main Street has faded. The stores are empty. The buildings are starting to crumble, and there is a sense that it is not going to change anytime soon. And always, *always*, there is a Dollar General or a Family Dollar somewhere on the outskirts of town. These are newly built and owned by a company located far, far away. If a town is large enough, they might even have one of each.

When you look at a map of voting patterns by county, there are no blue states and red states, there are blue urban counties and red rural counties. Whether a state is blue or red is based on the dominant force in the state.

Good thing we've still got politics in Texas—finest form of free entertainment ever invented.

—MOLLY IVINS

For Texas, it is rural. But it is changing. From 2010 to 2020, the fourteen urban counties have grown by 19.6%. They compromise 65% of the state's population, and in the 2022 governor's election, they voted 53.3% for the Democratic candidate Beto O'Rourke. Urban counties are defined by the US Census as greater than 500 people per square mile, rural is one hundred people per square mile or greater, and ultra-rural is fewer than one hundred people per square mile.

In Texas, there are thirty-two rural counties by the US Census definition. They had 15.1% growth between 2010 and 2020 and now have 20.1% of the population of Texas. In the 2022 election, they went for the Republican incumbent governor by 66.6%. Most of these "rural" counties are next to urban counties and experience growth from their proximity to the urban centers. Hays County is adjacent to the capital of Texas, Austin, which is in Travis County. Hays experienced 53.4% growth in the same ten years that Travis County grew by 26%. Several of the other rural counties next to urban areas had over 30% growth.

But get into the real rural areas of Texas, and the numbers change. Ultra-rural counties saw only 2.8% growth between 2010 and 2020 when the state experienced 15.9% growth. It gets even worse when you consider that the true positive growth was only in sixty-nine (33%) of the 208 ultra-rural counties and the remainder of the counties lost population. Over half of all Texas counties lost population while the state grew by almost 16%, and except for three rural counties, they were all ultra-rural. But they voted Republican in a big way. In 2020, they voted 74.1% for Trump as opposed to 24.8% for Biden, but in 2022, they voted 77.4% for Greg Abbott, securing his reelection as governor of Texas.

The divide in America is too often portrayed as Republican versus Democrat, conservative versus liberal, but it actually breaks down into urban versus rural, and rural is losing the demographic and economic battle. We need to understand why. There are a lot of questions why rural areas vote Republican when Republican policies actually hurt rural voters, and the reason is simple. Rural voters are farmers, ranchers, and small-town folks who have been fed a steady diet of doom-and-gloom news to make them distrust Democrats. Who is coming for your guns? Did you see all the illegal immigrants pouring across the border? Did you hear about the decline of white protestant America? Did you hear how we will be a minority soon? They have been told what's wrong and how it is getting worse, and when they look around their ultra-rural county, they have reason to believe it.

The decline in rural America is real, and it starts with economics. Draw a box around any ultra-rural county and start looking at where the money flows. Remember the dollar store on the outskirts? It replaced the stores owned by locals. A multi-billion-dollar corporation located out-of-state owns it, ultimately owned by wealthy stockholders. The products are mostly from foreign sources, even vegetables are sourced from produce companies that aren't in the county. Nothing is made in the county and wages paid to local workers are likely just above minimum wage. Everyone has a cellphone, none were made in America, and they all send money out of the county to corporations in one form or another. If the county is lucky, the crews who maintain the cell towers are local, but most likely they live in a larger county nearby.

Electricity? Everyone uses it, no one in the county owns any of the infrastructure, and maybe a power plant is located within the county. That could mean some well-paying jobs, but it could also be a large coal power plant not only producing electricity, but also producing pollutants. Multiple studies have shown that cases of asthma are higher in the vicinity of coal power plants, and a recent study in Louisville, Kentucky, from 2013 to 2016 looked at the impact of four large coal plants on asthma patients.[110] The study estimated that in the region surrounding Louisville, shutting down plants or installing exhaust scrubbers resulted in a 32% reduction in asthma inhaler use and 400 fewer hospitalizations or emergency room visits per year. Short of a power plant in the county, electricity is a net negative to the economy—money flowing out, but none flowing in.

In Texas, the state poverty rate is 13.6%, but in ultra-rural counties, it is 16.3%. That is average, but in the counties with a declining population, it is 16.9%. And that is an average with a huge variation from 30.5% poverty in the worst county (Willacy County near Brownsville) to 7.1% in the best (Loving County, presently enjoying revenue from the Permian Basin oil fields).

So where is the money flowing into the counties coming from? Farming and ranching are the mainstays of rural economies, but they also exploit resources. Oil revenues, mining, wind farms, and other localized industries are the drivers in that large variation in poverty rates. But remember counties with high levels of resource exploitation, especially oil revenues, are not part of the declining population group. Some of the ultra-rural counties with the largest percentage of growth are in the Permian oil basin. The ultra-rural counties of Andrews, Ector, Midland, Gaines, and Sterling have seen over 20% growth thanks to the Permian oil activity, and their average income per capita is higher than the state average. Contrast that with the shrinking counties where income per capita is $7,000 less than the state average.

The other issue is demographics. In 2010, these declining counties were 55% white and 8.5% Black. By 2020, that had declined to 51% white and 8% Black. The gains? Hispanics and "other." The Texas legislature tracks "other" as anyone not fitting into the basic buckets. Texas tracks population referring to Anglo, which is anyone on the US Census who checked white only and did not check Hispanic, a very limited focused group. Check white and Black on the US Census, then Texas considers you Black. Check Hispanic and any other race, and you are Hispanic. Check Asian and any other race, and Texas considers you Asian. And yes, there is an overlap between Hispanic and Black. To be Anglo, you must have only checked white on the US Census and nothing else. That seems like a throwback to the Jim Crow days, but that is how race has been framed in Texas. When the state draws up election districts, they report out based on those five simple groups—Anglo, Black, Asian, Hispanic, and Other. And before we begin to criticize Texas, these same basic rules are pretty much used by every other state and the US census.

The simple truth for Texas is that Anglo has been in decline. Overall, Anglo has declined by 5.6% from 2010 to 2020, and overall, in ultra-rural Texas, there is a 4.2% decline. But that is by percent. When you look at the actual numbers, it is even starker. For the declining counties, the population of Anglos has fallen by 134,025. In other ultra-rural counties, it has risen slightly, but for the declining counties, Anglos has been offset by an increase in Hispanics (15,562) and an increase in Other (35,120). For the ultra-rural counties where population hasn't declined, the bulk of the growth is Hispanic and Other.

If you start telling rural Texas that white protestants are becoming a minority, guess what? They will believe you, because they see it. Tell them that climate change is going to harm them, and think back to Chapter 3, when we showed that climate change impacts urban counties far more than ultra-rural counties. They aren't seeing that as much as they are seeing the boarded-up buildings and a declining Anglo population. Climate change is some vague remote threat. The changing economics and demographics in their hometowns are right in front of them.

And what are the implications? Well, first, rural America is going to respond to a promise of a return to the 1950s, 1960s, and 1970s, when rural towns were not turning into ghost towns. When local stores were owned and operated by local people and sold goods to the local population, rather than a dollar store and Amazon delivery trucks driving through on a weekly basis. In the rush to globalization, rural America got left behind while urban areas reaped the rewards. This is shown in the demographic data, where urban counties in Texas have an average income per capita of

$56,860, which is $4,000 above the state average, as opposed to ultra-rural counties where the income average is $10,000 less than the state average.

The loss of rural small farmers doomed the Roman Republic to change into a Roman Empire that eventually led to the collapse of the Roman civilization. What are the implications of the decline of small-town America and its ultra-rural counties to the American civilization? This is an area without an easy answer. The data clearly shows that rural areas are in decline, but what to do that doesn't involve increasing the national debt or massive government overreach?

The first action item is change local news. Many small-town newspapers have failed or are barely hanging on. The rise of cable and internet news has been a decimating force against local journalism, and local governments rarely get the coverage to keep them honest. Every high school should have a journalism class and get some support to allow them to cover and monitor their local school board. Local newspapers should be able to leverage high schools and get some support as well. The exact details need to be worked out, but a lack of monitoring and accountability of local governments is not healthy for the American democracy. Funding should be via a charge to internet content providers. This has been done in Australia and other countries based on the idea that Google, Microsoft, Apple, and others are using local journalism on their websites without providing payment to the journalists. What is needed is a fee to companies using these journalists' results to fund the local journalists to help them generate the stories that the internet sites then republish.

The second is taxes on corporations, as was discussed in the chapter on taxes. Small-town ownership of a business should be cheaper than a distant corporation owning the same business, and then you might see the revitalization of small-town Main Street.

The third is rural healthcare. Small hospitals are closing or being absorbed by corporate hospitals. This presents a challenge of loss of coverage in rural areas. The solution is to work to set up non-profit hospitals that could be tied to medical schools as teaching institutions. The chapter on education discussed scholarships for nurses and doctors and requiring them to work for two to four years in a rural setting. This would go a long way to helping restaff rural medical requirements. There may be tax relief or some incentive to encourage the development of teaching facilities. The patients would get low-cost care with the understanding that they are teaching facilities, and malpractice lawsuits would have a stricter bar to overcome.

And finally, the fourth item is the basic income discussed in a previous chapter. Any funds provided to the rural poor are going to not only help them survive but be an infusion of cash into rural economies that could desperately use it.

The hard truth is that just like the ancient Roman Republic, we have not been investing in our small towns, but allowing corporate America to plunder them. The globalization that has provided a massive benefit to urban areas has not trickled down to small towns, and as a result, too many of them are dying. It is now time to reverse that course and invest in the small towns that not only support the agriculture and industries in our rural regions but have formed the backbone of America's democracy.

44.
The American Duopoly

I'm neither party. The political duopoly is at the heart of why we are partisan. When you play us versus them, it gets worse. Want to see it get better? Take the money away from both parties.

—Mark Cuban

The middle of America's political spectrum finds itself standing outside two giant tents and unsure to which they belong. This is surprising because in a democracy, the government doesn't set policy; it is the society acting through the political parties who sets policy by electing representatives and expressing its views via the popular vote. In a single-party government, the party has divorced itself from the society and acts directly to set policy, taking control away from the society. The United States has always had a primarily two-party system, which does not divorce the party from society, but limit society's choices. At times, political parties have adjusted their views to conform to change in society, and in a few extreme examples, some political parties have collapsed. We view the Democrats and Republicans as the only options, but it needs to be pointed out that Republicans only formed in 1854, just prior to the American Civil War. The Federalists were founded by Alexander Hamilton at the start of the democracy and faded out of existence in the early 1800s. The Whigs replaced them to a great extent and were in turn replaced by the Republicans as the Whigs faded into history. At the present moment, we are seeing a shift in both parties, which is having a major impact on the Republican party as it deals with the aftermath of Donald Trump. Even the Democrats are not without problems as the progressive and moderate wings find themselves at odds over issues and which direction to take the country in.

At the core of this is the gerrymandering that has changed the fundamental alignment of elections. Politicians no longer concern themselves with the general election when they have custom-drawn districts that lean one way or the other. The problem becomes the primary when they compete for votes from the ultra-party members. Wonder why the moderates in both parties have faded away? It's the primaries. A moderate candidate can't compete in a landscape that favors the radical. And in the chapter on gerrymandering, we discussed how to reduce it, but the party in power has no desire to change the rules when they feel it favors them.

The solution to a two-party stalemate with increased radicalization and a larger number of Americans now identifying as moderate/centrist independents? Form a third party. Those four words should send a shudder through the existing party leadership on both sides of the divided government. And in response, some states have adopted, or are considering adopting, laws to make it even harder for third parties to get ballot access.

Ballot access is a party's ability to put their candidates on the November general election ballot. And the variation in requirements from state to state is massive. Texas, one of the hardest, requires that candidates for the third party register in November a year before the general election and before the party has even qualified for ballot access. Then as the primaries for the major parties conclude in March, the third party needs to conduct precinct meetings and gather signatures of registered voters who did not vote in the primaries. And oh, by the way, they only have seventy-five days, and they need to gather enough valid signatures to constitute 1% of all votes cast in the last governor's general election. For 2024, that would be 81,300 signatures of valid voters who will not have voted in either the Republican or Democratic primary. Those are just the rules for Texas and each state does it differently.

This is not advocating for open ballot access. But it is advocating for a better and fairer way to allow more voices to appear on the general ballot. That creates a bit of fear in political circles since some states don't allow runoffs and elect the winner of the general election with the most votes. Texas, with all statewide offices being single-winner with plurality, had a clear example of that in the 2006 governor's race. A four-way field was won by Rick Perry with only 39% of the vote. Think about that for a moment: 61% of the voters chose someone else, but Rick Perry was declared governor. So why not have runoffs?

Runoffs cost money. In the 2022 primaries, Texas spent more than $6 million on conducting runoffs for just the primary elections. That isn't what the candidates spent; it is what the state spent to hold the elections. Not only that, but runoffs never have the turnout that the primary election had. In the case of the Republican attorney general race, the primary had 1,927,457 voters, but the runoff only had 755,836 voters, fewer than 40% of the voters in the primary participated in the runoff.

But there is another option: ranked-choice voting (RCV). With RCV, instead of casting just one vote, you rank the candidates from your first choice to your last choice. You can then conduct an instant runoff. If no candidate gets over 50%, then the candidate with the lowest tally is eliminated and his voters then move to their second choices, and the votes are retallied. On the second round, if no candidate has gotten over 50%, then once again, the candidate with the lowest tally is eliminated and his voters move to their second choices, or third choices if they had voted for the candidate eliminated in the first round. This process keeps repeating until one candidate has more than 50% of the votes that are still in play.

So, what are the disadvantages? Well, the Foundation for Government Accountability (thefga.org) claims that it's confusing, will lower voter turnout, and create mistrust in elections. The problem is while they reference a 1995 to 2011 study of San Francisco elections that indicated an 8% decline compared with non-RCV elections to justify their claim of lower voter turnout, it doesn't correlate with actual results in Alaska and Maine, which use RCV for congressional elections.

Alaska's elections in 2022 were the first time they used RCV, and they saw a 25% decline in voting compared with 2020. But 2020 was the Trump versus Biden mega-battle, and that 25% drop is actually a lot better than the average 32% drop in voting that all thirteen small states (one or two House representatives) averaged. In fact, Maine, which used RCV voting in 2020 and 2022, only saw an 18% drop in voting compared with 2020. The facts are that looking at Maine and Alaska and comparing them with the other eleven small states, there is no proof of voter drop-off, and in fact, it looks like an improvement.

As for the reports of confusion, there really hasn't been any.

The other argument that FGA makes (as well as a number of other critics that have come out against RCV) is that some ballots are "exhausted" and that disenfranchises the voters. That is a weird argument, because an "exhausted" ballot is one where the voter's choices all get knocked out for having too few votes, and then their ballot doesn't have any of the remaining choices on it. But think about what that means. Even if they had a single ballot, then their choice wouldn't have won the plurality and wasn't going to make a runoff. This actually gives a lot more ballots the potential to count toward deciding the race.

For the Alaska US Senate race in 2022, 36,808 ballots in the first round were not for the top two candidates. But 20,571 of those ballots had picked Lisa Murkowski as their alternative between the 2,224 that picked Kelly Tshibaka, putting Murkowski up to 53.7% and winning the election. Only 9,163 ballots didn't pick Murkowski or Tshibaka, 3.5% of all ballots cast. That's actually a better result than kicking this over to a runoff, where only a fraction of the voters shows up. And if it had been a plurality contest, Murkowski would have won outright. The Alaska race for the US House was the same way, where Peltola would have won the plurality outright and did win in the third round of RCV with only 14,855 ballots being exhausted (5.6% of total votes) out of the 67,504 votes cast for the third- and fourth-place candidates that wouldn't have counted in a runoff or plurality scenario.

The fact is that Foundation for Government Accountability is a Florida-based think tank supported by Republican donors and rightwing PACs, as are all the other voices against RCV. The Association of Mature American Citizens (amac.us), which presents itself as a conservative-leaning alternative to AARP (American Association of Retired Persons) indicated that

ranked choice voting (RCV) is a confusing, chaotic election "reform" being pushed by mega-liberal political donors and other activists that would fundamentally change the election process, effectively disenfranchise voters, and allow marginal candidates not supported by a majority of voters to be elected.

The conservatives view RCV as a threat after the Alaskan elections where the Trump-endorsed candidates lost, and they are now actively working against RCV. Multiple Republican-dominated state legislatures are trying to pass bills that would ban the use of RCV for any elections. There is even a move in Alaska to try and get RCV repealed. Why? Because if voters get used to using RCV in local elections, they might want them for statewide or even national elections, and conservatives fear that RCV could spell disaster for them.

The simple reason that RCV saves governments millions of dollars to conduct runoff elections should be reason enough to use it. But this negates the argument of the spoiler effect that third-party candidates have historically presented, which has been used to lock third parties out of the political process. Political scientists see ranked-choice voting as a way to move America away from the ideological edges and give voters the ability to elect moderate candidates.

So, what about a third party? Is it time? Maybe. Studies are showing more people have abandoned the two major parties and are declaring themselves to be independents. But that doesn't mean that a third party can be viable.

In July 2022, the Forward Party organized by Andrew Yang merged with the Renew America Movement (RAM) and Serve America Movement (SAM). All three were independent moderate parties. I had joined SAM in 2021 and am now helping the Forward Party in Texas, so I do have a bias in this. My view is that with both the major parties on the far ends of the political spectrum, it should be time.

Conclusion: The Hope for Moderation

Moderation (noun): the avoidance of excess or extremes, especially in one's behavior or political opinions.

—Oxford Languages

Political liberty is to be found only in moderate governments.

—Baron de Montesquieu

The virtue of justice consists in moderation, as regulated by wisdom.

—Aristotle

I never smoke to excess—that is, I smoke in moderation, only one cigar at a time.

—Mark Twain

45.
Religious Tolerance

We have tolerance, respect, and equality in our written laws but not in the hearts of some of our people.

—Ruby Bridges

There can be no government policy on religion, because our Founding Fathers specifically said in the First Amendment that "Congress shall make no law respecting an establishment of religion, or prohibiting the free exercise thereof." That has become a bedrock of American law and has led to the policy of separation of church and state. In a 2011 article in *Forbes*, Bill Flax identified one of the original sources of that concept.

The phrase "separation of church and state" was initially coined by Baptists striving for religious toleration in Virginia, whose official state religion was then Anglican (Episcopalian). Baptists thought government limitations against religion illegitimate. James Madison and Thomas Jefferson championed their cause.[111]

But he goes on to argue that this does not prevent public displays of faith at government facilities. But when do these displays move from an expression of society's views to one of government sponsorship? Because the Founding Fathers understood a basic concept, a state-dictated religion will always be incompatible with a democratic republic.

And yet we have groups proclaiming that religion has been too excluded from government and the act of removing religious symbols based on courts' orders is massive judicial overreach. Due to these actions, I had considered putting religion as one of the hot topics within American society, but in some ways it isn't. Most people in America believe in God. A Gallup poll in 2011 asked the simple question, "Do you believe in God?" And 92% of Americans surveyed said yes. It is how we believe in God that divides us, and it is how this has changed over the years that has caused religion to become an issue. In 1948, 68% of Americans were Protestant, 22% were Catholic, and 4% were Jewish. Jump to 2019 and the numbers are completely different: 43% Protestant, 20% Catholic, 2% Jewish, 2% Mormon, and 26% Unaffiliated. Mormon and unaffiliated weren't even part of the question in 1948. Of the 43% Protestants, one third of them identify as evangelical. Christianity, which is defined by some as Protestants, Catholics, and Mormons when lumped together, was 90% of the country in 1948; it is now 65%. For white protestants, the last seventy years have brought a decrease in their numbers.

And this is the problem. White protestants have become a majority of those calling for a return to an older America, an America where they were the predominant voice, which they imagine as a better time. The problem with that is while it was a better time for white protestants, it certainly wasn't for minorities. And that call for older times ignores the advances that America has made since the 1950s in everything—science, medicine, culture, and diversity. It ignores the demographic changes that have been reflected in some of the problems we are facing. It is also unrealistic. The changes of the past can't be reversed without great upheaval. To return to a time that was better for the conservative majority begs the question: what price would the minorities have to pay? Because in some areas, those minorities are now starting to make up the bulk of the population.

Which is why protestant symbolism no longer reflects the composition of the country. Does that mean we need to tear down these symbols? No, for two reasons. First with a massive debt, not one federal dollar should be spent removing or rebuilding anything. In oil field terms, "if it ain't broke, don't fix it." If some people are offended by walking by Christian symbols, we ask their tolerance because for the second reason, this was and still is a predominantly Christian country. It has, in fact, the largest Protestant population of any country in the world. It is the history of this country, and we should leave it as such. However, that doesn't mean that federal or state dollars should be used to promote any specific religion, but if the local fire department puts up a Christmas tree or even a manger scene, then let them. There is a gray line somewhere for when a religious display becomes an overtly offensive symbol to other religions. But rather than ban all religious displays, let's try and find that line, maybe even allow some displays to brush up against it rather than forbidding all displays. If you put up a manger scene or a tree, then you really can't object to, nor disrespect, a menorah or any symbol of an accepted religion. It is time to stop seeing every small perceived slight as requiring remedy and ask all people for tolerance and moderation. And moderation works both ways.

46.
The Big Problems

If we were really tough on crime, we'd try to save our children from the desperation and deprivation that leave them primed for a life of crime.

—Carrie P. Meek

The major problems facing America are greed, overpopulation, and poverty. In the preceding chapters, we have looked at America's symptoms and in each one you can find one of these root problems influencing the issue. In some cases, they are the total root and cause. And while the solutions proposed in the previous chapters are a way to get the country on the right track, there still needs to be a discussion of the three core problems.

It is greed to do all the talking but not to want to listen at all.

—Democritus

The greed of the wealthy wanting to keep and increase their wealth coupled with the greed of politicians and social leaders who want to maintain and increase their power have always existed. If you are on top, you want to stay on top. But periods where the wealthy have dominated society are usually followed by periods of revolution where the oppressed rise up and turn the tables. The wealthy can fret over "redistribution of wealth," but a little redistribution tends to preclude the revolution of wealth that has happened so many times. And America may be due for some redistribution. The Gini coefficient that measures wealth inequality has been on the rise. The last time it was this high was in the 1920s, just before the Great Depression.

The Great Depression started in 1929 when the US stock market collapsed, and a global depression settled into place that caused massive unemployment and hardship. The Republican party had been in charge since 1920, and their policies of low taxes for the wealthy and corporations had been sold as helping spark the economy after the recession that marked America's transition from the wartime economy of World War I to a peacetime economy. The "Roaring Twenties" marked massive growth in new technologies—cars, planes, electricity, and other developments sparked an economic boom. But that boom was limited to the wealthy and upper classes. Republican policies against unions, agriculture, and social welfare meant that the poor, the farmers, and the factory workers were excluded from the prosperity, hence the very high inequality expressed by the Gini coefficient of the 1920s. At the same time, the Ku Klux Klan had re-emerged, lynching was on the rise, and a number of violent events occurred. The burning of Black Wall Street in Tulsa in 1921 is but one example, and there were many others that show a pattern of rising social tensions not only among African Americans, but more generally.

This was also the era of Prohibition, where a moral minority was able to outlaw alcohol for the entire country. The Eighteenth Amendment that banned all alcohol was ratified in 1919 and was the law of the land from January 1920 until December 1933 when the Twenty-first Amendment was ratified overturning the Eighteenth.

Instead of sparking a "tide that floats all boats," Republican economic policies sparked a wealthy upper class that invested in stocks and in some cases bought them on credit. This created a Wall Street bubble that burst on Black Tuesday, October 29, 1929. Sixteen million shares traded hands on a single day as the Dow Jones Industrial lost 12% of its value after it had lost 13% the previous day. Over the next two years, the Dow Jones continued to decline, losing 89.2% of its value.

A rising inequality of income, a minority imposing a moral position on the rest of society combined with social tensions to form a pattern that we are seeing right now as well. And as has been said, "History doesn't repeat itself, but it often rhymes."

Another data point to look at is the pay for chief executive officers compared to typical workers. A report by the Economic Policy Institute in October 2022 indicated that the pay gap has been increasing.[112] In 1965, CEOs made twenty times what typical workers made. By 2021 that ratio increased to almost 400 times a typical worker. The report showed that the worker-to-CEO gap had been increasing, and by the 1990s it reached a point where CEO compensation was over sixty times that of a typical worker, but in the 2000s, the compensation packages exploded, and CEO compensation jumped to over 300 times that of workers. When you look at the largest US companies, the gap widens even further.

And this is not just about financial greed. As other the chapters discussed, we use too many resources in throwaway products and generate too much trash. As resources become scarce, this will become a serious problem and we only need to look at those countries that are less wealthy, less blessed with natural resources, to see the problems that are on the horizon.

The chapter on Basic Income may have upset some people as "giving away hard-earned money." But a very small tax on stagnant money that the wealthy are sitting on in stocks, bank accounts, and retirement funds provided to the poor who actually spend it, is a way to improve the economy, and far more effective than tax breaks to the wealthy. A basic income not only provides a safety net for everyone in poverty who is willing to make the effort to access it, but it also benefits the economy.

In the discussion of politics, we looked at two fundamental political philosophies of equal opportunity and protection of property. Since the 1980s when Reagan dramatically lowered taxes on the wealthy, and the 2000s when manufacturing jobs went overseas to provide cheaper imports to America, the pendulum has swung toward protecting property rights. The fear is a major swing to the other side, to equal opportunity for all. It is seen as a massive "redistribution of wealth," a phrase some progressives can't help but blurt out at times. But again, the solution is not at one extreme or the other; it is the moderation of finding the balance where people can rise up on their effort, while those who have earned more and have more are not threatened. A single waitress could once work one job and be able to put her children through college. When some people are talking about a return to a better time, that is one of the things they are longing for. While that may have been true in the past, today there are too many examples of people working multiple jobs and barely keeping their heads above water to ignore. Opportunities to rise have become limited.

Greed is hard to address. It is painfully obvious when we have billionaires sailing by on mega-yachts and the working middle class can't find affordable or effective childcare so they can work. And while some can proclaim "tax the rich," it does no good if the rich only turn around and raise the rents of their low-income rental properties. The challenge comes back to finding the balance between protecting the property rights of everyone, but also taxing the wealthy to help lessen the inequality and provide the poor and middle class with an opportunity.

The wealthy need to come to a decision. They can participate in America and help improve the lives of all Americans, or they can sail off into the sunset before the inequality and the national debt reach a breaking point.

Futurists don't consider overpopulation one of the issues of the future. They consider it the issue of the future.

—DAN BROWN

The United States has grown from a country of 176 million in 1960 to over 338 million in 2022. That pales in comparison to the worldwide population that increased from less than 3 billion in 1960 to over 8 billion today. That growth drives demands for energy, housing, and food. Environmentalists can argue that we need to stop fossil fuels to save the planet, but that demand has been driven by a massive increase in population that can't be easily satisfied by renewables. In the chapter on agriculture, we also looked at what happened when the cheap fertilizer supplied by fossil fuels goes away. Every alternative on the horizon uses more energy and is costlier than the cheap fertilizer we use. People talk of hydrogen and how it can make ammonia to feed that fertilizer process, but hydrogen has to be made from something and more studies show that it can be very energy inefficient. Energy, housing, and food usage become critical when you look at how many people internationally live in poverty and lack those necessities and how much of an increase would be needed to provide them with even a basic American standard of living.

All the solutions presently being looked at are premised on today's population size. The United Nations Department of Economic and Social Affairs has a population division that tries to predict world population. They have multiple scenarios, but most of them show us reaching 9 billion sometime before 2040. Most scenarios show population plateauing somewhere around 10 billion before beginning to decline. In the chapter on abortion, I pointed out that the majority of this growth is in countries that do not allow abortion. Those same places are plagued with poverty and poor education and are now indicating that they will try and reduce their population growth, but few are looking at legalizing abortion.

This book makes no attempt to fix things on a worldwide scale. It is intended to help the American public see through the misinformation from both the left and right to understand the issues at the heart of our problems. So, we need to focus on what's happening in America. And that has to bring us back to abortion, immigration, and efforts to stem population growth. America has welcomed immigrants as part of its identity, but that has to change if we are going to address our problems. An influx of asylum seekers, outlawing abortion, and an existing trend where most of our children are born in or near poverty, means we are not in a position where population growth is beneficial. As it stands with birth and immigration, more growth is happening below the poverty line than above and to change that, we need to seriously consider the recommendations in the chapters on immigration and abortion.

There is no easy solution to all of America's problems, but it becomes harder if our population begins to skyrocket. And the continued use of the Helms Amendment that blocks any US foreign aid being used for abortion means we are not doing anything to address the worldwide population crisis. While some people find abortion to be wrong, the massive population growth in poor countries around the world is placing millions of children in poverty, and it is the biggest factor driving people to immigrate legally and illegally to the United States.

In a country well governed, poverty is something to be ashamed of. In a country badly governed, wealth is something to be ashamed of.
—CONFUCIUS

Poverty has been the abiding problem in America. In the 1950s, 22% of Americans, almost 40 million, were living in poverty. The rate fell through the 1950s and 1960s, reaching 11.1% in 1973. Since that point, it has fluttered between 11% and 15%, reaching an all-time low in 2019 at 10.5% before COVID caused it to rise. The data for 2021 indicates it has risen to 12.8%; that is over 41 million Americans, more than were living in poverty in the 1950s.

But that is the poverty rate for adults; for children it runs higher. That is one of the horrible truths about poverty. Using the 2020 census data, 15.5 million children lived below the poverty line, 21.4% of all children eighteen and younger. And that is the poverty line, when you look at families making less than twice the poverty line, it is almost 50% of our children.

The war on poverty, much like the war on drugs, has been fought with lots of pomp, but very few hard victories. Government safety nets have been implemented, but they are focused on certain areas, sometimes require massive paperwork to enroll, and far too often miss the mark.

At the same time, check cashing and payday loan stores can be found in every poor neighborhood and most small towns. In his book *Poverty by America*, Matthew Desmond details how the payday loan business drains the finances of the working poor. And why? Because they live paycheck to paycheck and can't afford to use regular banks. Even when they do, Desmond documents how overdrafts and account fees fall predominantly on the lower-income clients.

And why have we allowed the poor to be preyed upon, to be nickel-and-dimed with fees and costs that keep them in poverty? Because our views of poverty vary so greatly. Just as the major aspects of America, society, economy, and government, all have different facets and conflicting needs, so do the poor. And which of those aspects leaps to mind when you think of the poor? Is it the struggling families who can't seem to make enough money? Is it the violent crime constantly shown to us on television? Is it the homeless man on the corner pleading for just a few dollars so he can have a meal? And the truth is they are all valid. The poor encompass everything from the homeless who have almost given up to the struggling families trying to find a path out of poverty, to the criminals hardened by poverty into the inmates of our most violent prisons.

Poverty is a burden to America. It is what gives us worse health outcomes, worse educational scores, and higher rates of crime than the other wealthy industrialized countries. Until we deal with poverty in a way that makes a difference, that will not change. Focusing on policing, prisons, public schools, and healthcare is only addressing the consequences of the underlying poverty. Until we address the root cause and reduce poverty, we are applying band-aids on festering wounds without truly healing them.

The idea of a basic income is going to offend many, but let's consider the other choices. Raising the minimum wage to the point where workers can actually have a good life would result in a massive cost increase in the cheap products that many have come to depend on. It also has the potential to crush a lot of small businesses, making the job market even tougher. But taxing the wealthy to both balance the federal budget and supplement the incomes of the poor and working poor would keep those product costs low while providing a better outcome for the workers. Yes, it is a limited redistribution of wealth, but it stops the slide into inequality that we have been experiencing for the last fifty years.

In industrial systems, there are recycle lines, where the high-pressure outlets of pumps send a fraction of the outflow back to the inlet to help stabilize the operation of the pump. It is a basic tool in any engineering design because it makes the system more stable and safer. A program of basic income is just like that recycle line. It would take a fraction of the money from the wealthy side of American society and bring it into the poor side of American society where the labor force exists that drives the economy to make the wealth. To make that poor part of the society more stable is to make the country safer and stronger.

The other option is to continue down the same road with the poor getting poorer and the wealthy getting wealthier until the system breaks. Having investigated the results of a recycle line not working and a pump having ripped itself apart, I can tell you that isn't going to be pretty or have a happy ending.

47.
Final Words

> Kindness is the language which the deaf can hear and the blind can see.
>
> —Mark Twain

This book started because of questions—questions about climate change, immigration, racism, federal debt, and so many other issues. Questions that weren't being answered, at least not in the way that I wanted, in terms of showing the data and analysis, so people could reach their own conclusions. The months before the November 2020 elections were filled with political commercials, online commentaries, and social media posts that told us what to feel, how to vote, but never told us why. And that led me to start some research.

It started with climate change and digging through all the weather data that NOAA keeps on file. It then led to the culture wars and trying to find out what was really going on. And it led into the national debt and the calculations of what interest payments would do as the Federal Reserve was raising interest rates in 2022. It led through reviewing the federal budget and how we have run deficits for the last twenty years and have little hope of a balanced budget in the future. And it led me to start writing. Documenting what at times felt like a journey down a very strange rabbit hole into facts and history that don't get taught in public school.

The good news is as a country, we have been in crisis before, critical periods with a very eerie similarity with today. The years before the Civil War are taught as if Abraham Lincoln was destined to be president; the truth is far more complicated. He entered the Republican convention in 1860 as the underdog and somehow emerged a leader who was critical to keeping the Union from being permanently rent asunder. What would have happened if an assassin's bullet hadn't derailed his plans for Reconstruction? Where would we be now?

It's an interesting puzzle for historians to kick around, but as an engineer, the question is what we do now.

We find ourselves in a transition. Not only in a transition on energy, climate, and culture, but decisions in past decades have run up deficits, allowed manufacturing jobs to move overseas, reduced taxes, and established social safety nets without the means to permanently fund them. The resulting tensions have changed the nation and are now coming to the forefront and confronting us.

We have extremes on either sides espousing hate and anger at the other extreme, which is strangling the very discussion that is needed. And that is not a new thing. At each of the past critical points, extremes have formed and called the other side the same epithets that we still use today. Extremes never solved the crisis, but only prolonged it. A victory by a progressive urban faction or a conservative rural faction will not save America, but only increase the risk to our democracy. Our path out of this crisis does not exist in the hate or anger, or in one side winning. It exists in having discussions between all Americans and finding solutions.

The point of this book was to strip away the radical propaganda, dig up the undiscussed history and facts, and get down to the core issues to see if there were solutions. And there are solutions. Solutions that everyone might not agree on, but solutions that form the basis for further discussion. And it is that discussion that we are lacking, a discussion tempered with kindness and regard for the opposite side. But the only way to get that is to stop electing politicians who embrace extreme views, who launch verbal attacks without a shred of evidence, and who won't have the discussions that we need. They won't be made by ideologues appealing to an extreme base of voters. They will only be made by leaders acting in the best interest of the country with moderation, wisdom, and tolerance. Those are the leaders we need to elect to start the process of fixing America.

APPENDIX

Guns, Poverty, Suicide, and Homicide Internationally

The comments about guns and poverty were the result of analysis that was done with Wikipedia's pages on "List of countries by percentage of population living in poverty" and "List of countries by firearm-related death rate." These two tables were combined into an Excel spreadsheet and then sorted based on the number of homicides per 100,000 citizens and the number of suicides per 100,000. The two resulting tables are shown in the following pages. The two sets of poverty data were provided by the World Bank and the US Central Intelligence Agency (CIA). The CIA estimated the percentage of the country's population living below the country's own declared poverty line. Some countries have very low poverty lines, and it gives them a skewed result. Brazil is an example; according to their own poverty line, 4.6% of the population was living in poverty, but the World Bank puts the number at 26% living in poverty for Brazil, based on the World Bank's criteria. Georgia (ex-USSR not the US state) and Montenegro are also examples of this. The World Bank's data set estimates the people living in the country on less than three different levels: $1.90/day, $3.20/day, and $5.50/day. The problem with the World Bank data is that it doesn't take into account when the cost of living in a country could be very low or very high. One example is the United States where fewer than 1.7% lives on less than $5.50/day, but the CIA puts the poverty rates as 11.8%. This is why it is important to look at both the World Bank and CIA data for a better picture of poverty.

The results are interesting.

The first table was sorted for suicides per 100,000 citizens. The result was a correlation with the number of guns per citizen and the gun suicide rate, but one that doesn't hold with all countries. The United States, at 7.32 gun suicides per 100,000. has the highest rate and with 120 guns per hundred citizens, it has the most guns in the hands of private citizens in the world. The next six countries all have rates of gun suicide greater than 2.0 and greater than twenty guns per hundred citizens, except Croatia which has thirteen per hundred, but Croatia has the highest poverty rate of the six countries. Countries with fewer than five guns per hundred citizens have the lowest gun suicide rates. The exception is Peru with high poverty and almost as many gun homicides as the United States, but a low suicide rate, something that needs to be checked to ensure that gun suicides aren't being under-reported. So, with a few exceptions, the more accessible guns are within the population, the more they are used for suicide. Five countries, all with very low gun ownership, did not report any gun suicide data.

While there is a correlation between the number of guns and gun suicides, it would take a massive reduction in the number of guns before any reduction in the number of gun suicides would be seen. And as was discussed in the chapter on guns, reducing access to guns would reduce the number of gun suicides, but would not reduce the number of overall suicides. Owning a gun doesn't provoke people to suicide, it just provides them with a means readily at hand for the suicide. Lacking a gun, the likely result is that a different method of suicide would be utilized.

Country	% Population living on less than $/day, by World Bank			% of Citizens in Poverty by CIA	Deaths from Guns per 100,000 citizens		Guns per 100 Citizens
	< $1.90	< $3.20	< $5.50		Total	Suicide	
Japan	0.20%	0.50%	1.00%	16.10%	0.02	0.01	0.60
Azerbaijan	0.00%	0.00%	8.20%	4.90%	0.23	0.01	3.60
Peru	2.60%	8.30%	22.10%	22.70%	3.72	0.01	18.80
Philippines	6.00%	18.70%	30.80%	21.60%	7.72	0.03	4.70
India	20.60%	22.50%	87.40%	21.90%	0.30	0.04	4.20
Romania	3.50%	7.00%	15.60%	22.40%	0.14	0.06	0.70
Kyrgyzstan	0.90%	15.50%	61.30%	32.10%	0.72	0.07	0.90
Zimbabwe	21.40%	47.20%	74.00%	72.30%	0.39	0.09	4.60
Poland	0.00%	0.10%	2.10%	17.60%	0.20	0.10	1.30
El Salvador	1.50%	9.70%	25.70%	32.70%	44.45	0.11	5.80
Panama	1.70%	5.20%	12.70%	23.00%	9.95	0.14	21.70
Venezuela	10.20%	17.80%	35.60%	19.70%	49.73	0.14	18.50
United Kingdom	0.20%	0.20%	0.70%	15.00%	0.20	0.16	12.80
Netherlands	0.00%	0.20%	0.50%	8.80%	0.42	0.24	2.60
Moldova	0.00%	0.90%	13.30%	9.60%	0.77	0.26	7.10
Guatemala	8.70%	24.20%	48.80%	59.30%	25.48	0.29	13.10
Nicaragua	3.20%	12.80%	34.80%	29.60%	4.68	0.30	7.70
Jamaica	1.70%	9.10%	29.70%	17.10%	35.22	0.31	8.10
Honduras	16.50%	30.00%	50.30%	29.60%	29.40	0.41	11.24
Spain	0.70%	0.70%	2.20%	21.10%	0.57	0.43	10.40
Israel	0.20%	0.70%	2.70%	22.00%	1.38	0.43	7.30
Mexico	1.70%	6.60%	23.00%	46.20%	17.00	0.46	15.00
Brazil	4.40%	3.70%	19.80%	4.20%	23.93	0.46	8.60
Chile	0.30%	0.70%	3.70%	14.40%	2.79	0.54	12.10
Ireland	0.20%	0.50%	0.70%	8.20%	0.87	0.62	7.20
North Macedonia	5.20%	9.70%	23.10%	21.50%	1.19	0.67	29.80
Colombia	4.10%	10.90%	27.80%	28.00%	20.38	0.69	10.10
Denmark	0.10%	0.20%	0.20%	13.40%	0.91	0.72	9.90
Italy	1.40%	1.30%	3.10%	29.90%	1.13	0.72	11.90

Country	% Population living on less than $/day, by World Bank			% of Citizens in Poverty by CIA	Deaths from Guns per 100,000 citizens		Guns per 100 Citizens
	< $1.90	< $3.20	< $5.50		Total	Suicide	
Bulgaria	1.30%	3.10%	7.50%	23.40%	1.51	0.73	6.20
Hungary	0.60%	0.60%	3.00%	14.90%	0.85	0.74	5.50
Costa Rica	1.40%	3.60%	10.90%	21.70%	7.59	0.90	10.00
Germany	0.00%	0.00%	0.20%	16.70%	1.04	0.91	32.00
Sweden	0.20%	0.20%	0.50%	15.00%	1.31	0.96	23.10
Slovakia	1.30%	1.30%	3.20%	12.30%	1.89	0.97	8.30
Greece	0.90%	0.70%	4.70%	36.00%	1.35	1.02	22.50
Paraguay	1.60%	5.90%	17.00%	22.20%	6.32	1.06	17.00
Belgium	0.10%	0.20%	0.30%	15.10%	1.40	1.09	6.86
Portugal	0.40%	0.30%	1.80%	19.00%	1.48	1.12	8.50
Latvia	0.70%	1.50%	4.00%	25.50%	1.86	1.16	19.00
Estonia	0.30%	0.30%	1.00%	21.10%	1.34	1.19	9.20
Canada	0.50%	0.50%	0.70%	9.40%	1.94	1.40	34.70
Czech Republic	0.00%	0.10%	0.40%	9.70%	1.64	1.43	16.30
France	0.00%	0.20%	0.20%	14.20%	2.33	1.64	14.96
Slovenia	0.00%	0.00%	0.10%	13.90%	1.91	1.71	13.50
South Africa	18.90%	37.60%	57.10%	16.60%	10.47	1.80	12.70
Serbia	5.50%	10%	20.30%	8.90%	3.23	2.15	37.82
Switzerland	0.00%	0.00%	0.00%	6.60%	2.64	2.32	41.20
Croatia	0.50%	1.20%	3.80%	19.50%	2.83	2.39	13.70
Austria	0.30%	0.40%	0.70%	3.00%	2.75	2.44	30.00
Uruguay	0.10%	0.40%	2.90%	9.70%	11.67	4.55	31.80
Montenegro	0.00%	0.80%	4.80%	8.60%	8.68	6.59	23.10
United States	1.00%	1.00%	1.70%	11.80%	12.21	7.32	120.50
Belarus	0.00%	0.00%	0.40%	5.70%	??	??	7.30
Bolivia	4.50%	4.10%	22.80%	38.60%	??	??	2.50
Ukraine	0.00%	0.40%	4.00%	3.80%	1.36	??	6.60
Georgia	4.50%	15.70%	42.90%	9.20%	6.39	??	7.30
Eswatini	42.00%	64.40%	82.00%	63.00%	37.16	??	6.40

TABLE 10. GUN SUICIDES

The following table was sorted for gun homicides, and sixteen countries had higher rates of gun homicides and they all had higher poverty by either the World Bank criteria or the CIA data provided, except for one, Uruguay. The range of gun ownership varied from the Philippines at 4.7 guns per hundred citizens to Uruguay at 31.8 guns per hundred citizens. There are a number of countries with less than 1.00 gun homicide per 100,000 citizens which have high levels of poverty, and all but three have very low levels of gun ownership with less than eight guns per hundred citizens. Again, there are a few countries that don't follow the pattern.

Uruguay has high gun ownership, 31.8, slightly lower poverty than the United States, but a gun homicide rate slightly higher than the United States. The Wikipedia data is eight years old at the writing of this; newer data from Latin American news outlets shows higher rates of gun homicides and gun ownership. The news outlets theorize that these are coupled with organized crime bleeding over the borders from Brazil.

The Philippines' high gun murder rate, but low gun ownership, breaks the pattern, but it has to be considered in light of state-sponsored vigilante killings that had been occurring at the time. The government's crackdown on drug dealers is the reason they are in the top homicide rate with very low gun ownership. Greece and Serbia are another exception in that they have high poverty, high gun ownership, but low rates of gun homicides. Greece has long had a reputation as having one of the lowest crime rates in Europe, but how that is created is not clear and it doesn't fit with the pattern seen in most countries. Except for these three, there is a general pattern. High rates of gun ownership in countries with low poverty don't have murder rates anywhere near that of the United States. Germany, Switzerland, Canada, and Austria are examples. High rates of poverty have high rates of gun homicides, except where gun ownership is very low, less than eight guns per hundred people.

Now only fifty-eight countries were considered, as many lacked data on gun homicides. Several had such low rates of poverty that neither the CIA nor the World Bank have provided data for them, even though they have high levels of gun ownership. Norway and New Zealand lacked poverty data, but each has more than thirty guns per hundred citizens and are likely in the same category as Germany, Switzerland, Canada, and Austria.

The conclusion is that reducing the quantity of guns won't have an impact on gun homicides except if it is brought down to very low levels, where over 90% of the guns would need to be confiscated, and that is just not feasible in the United States.

Country	% Population living on less than $/ day, by World Bank			% of Citizens in Poverty by CIA	Deaths from Guns per 100,000 citizens		Guns per 100 Citizens
	< $1.90	< $3.20	< $5.50		Total	Homicide	
Japan	0.20%	0.50%	1.00%	16.10%	0.02	0.00	0.60
Estonia	0.30%	0.30%	1.00%	21.10%	1.34	0.00	9.20
Slovenia	0.00%	0.00%	0.10%	13.90%	1.91	0.00	13.50
United Kingdom	0.20%	0.20%	0.70%	15.00%	0.20	0.02	12.80
Poland	0.00%	0.10%	2.10%	17.60%	0.20	0.03	1.30
Romania	3.50%	7.00%	15.60%	22.40%	0.14	0.04	0.70
Hungary	0.60%	0.60%	3.00%	14.90%	0.85	0.05	5.50
Germany	0.00%	0.00%	0.20%	16.70%	1.04	0.06	32.00
Spain	0.70%	0.70%	2.20%	21.10%	0.57	0.10	10.40
Czech Republic	0.00%	0.10%	0.40%	9.70%	1.64	0.10	16.30
Austria	0.30%	0.40%	0.70%	3.00%	2.75	0.10	30.00
France	0.00%	0.20%	0.20%	14.20%	2.33	0.12	14.96
Switzerland	0.00%	0.00%	0.00%	6.60%	2.64	0.13	41.20
Belarus	0.00%	0.00%	0.40%	5.70%	??	0.14	7.30
Netherlands	0.00%	0.20%	0.50%	8.80%	0.42	0.16	2.60
Denmark	0.10%	0.20%	0.20%	13.40%	0.91	0.18	9.90
Greece	0.90%	0.70%	4.70%	36.00%	1.35	0.19	22.50
Azerbaijan	0.00%	0.00%	8.20%	4.90%	0.23	0.20	3.60
Bulgaria	1.30%	3.10%	7.50%	23.40%	1.51	0.20	6.20
Ireland	0.20%	0.50%	0.70%	8.20%	0.87	0.21	7.20
Portugal	0.40%	0.30%	1.80%	19.00%	1.48	0.24	8.50
Belgium	0.10%	0.20%	0.30%	15.10%	1.40	0.25	6.86
Italy	1.40%	1.30%	3.10%	29.90%	1.13	0.29	11.90
India	20.60%	22.50%	87.40%	21.90%	0.30	0.30	4.20
Kyrgyzstan	0.90%	15.50%	61.30%	32.10%	0.72	0.30	0.90
Zimbabwe	21.40%	47.20%	74.00%	72.30%	0.39	0.30	4.60
Slovakia	1.30%	1.30%	3.20%	12.30%	1.89	0.30	8.30
Moldova	0.00%	0.90%	13.30%	9.60%	0.77	0.31	7.10
Croatia	0.50%	1.20%	3.80%	19.50%	2.83	0.35	13.70

Country	% Population living on less than $/ day, by World Bank			% of Citizens in Poverty by CIA	Deaths from Guns per 100,000 citizens		Guns per 100 Citizens
	< $1.90	< $3.20	< $5.50		Total	Homicide	
Sweden	0.20%	0.20%	0.50%	15.00%	1.31	0.40	23.10
Latvia	0.70%	1.50%	4.00%	25.50%	1.86	0.40	19.00
Israel	0.20%	0.70%	2.70%	22.00%	1.38	0.68	7.30
Canada	0.50%	0.50%	0.70%	9.40%	1.94	0.72	34.70
Serbia	5.50%	10%	20.30%	8.90%	3.23	0.72	37.82
Bolivia	4.50%	4.10%	22.80%	38.60%	??	0.74	2.50
North Macedonia	5.20%	9.70%	23.10%	21.50%	1.19	1.10	29.80
Ukraine	0.00%	0.40%	4.00%	3.80%	1.36	1.36	6.60
Chile	0.30%	0.70%	3.70%	14.40%	2.79	1.92	12.10
Montenegro	0.00%	0.80%	4.80%	8.60%	8.68	2.50	23.10
Peru	2.60%	8.30%	22.10%	22.70%	3.72	3.20	18.80
Nicaragua	3.20%	12.80%	34.80%	29.60%	4.68	3.70	7.70
United States	1.00%	1.00%	1.70%	11.80%	12.21	4.46	120.50
Uruguay	0.10%	0.40%	2.90%	9.70%	11.67	4.70	31.80
Paraguay	1.60%	5.90%	17.00%	22.20%	6.32	6.00	17.00
Georgia	4.50%	15.70%	42.90%	9.20%	6.39	6.39	7.30
Costa Rica	1.40%	3.60%	10.90%	21.70%	7.59	6.46	10.00
Philippines	6.00%	18.70%	30.80%	21.60%	7.72	7.62	4.70
Panama	1.70%	5.20%	12.70%	23.00%	9.95	9.30	21.70
South Africa	18.90%	37.60%	57.10%	16.60%	10.47	12.92	12.70
Mexico	1.70%	6.60%	23.00%	46.20%	17.00	16.50	15.00
Colombia	4.10%	10.90%	27.80%	28.00%	20.38	18.20	10.10
Guatemala	8.70%	24.20%	48.80%	59.30%	25.48	20.41	13.10
Brazil	4.40%	3.70%	19.80%	4.20%	23.93	22.91	8.60
Venezuela	10.20%	17.80%	35.60%	19.70%	49.73	26.48	18.50
Honduras	16.50%	30.00%	50.30%	29.60%	29.40	28.65	11.24
Eswatini	42.00%	64.40%	82.00%	63.00%	37.16	37.16	6.40
Jamaica	1.70%	9.10%	29.70%	17.10%	35.22	38.20	8.10
El Salvador	1.50%	9.70%	25.70%	32.70%	44.45	44.45	5.80

TABLE 11. GUN HOMICIDES

Guns, Poverty, and Individual States

The correlation between poverty and guns also holds true when individual states are compared against each other. The table on the next page is sorted based on the rate of gun deaths (homicide and suicide) per 100,000 adult citizens.[113] The key factor in the table is GDP per citizen, which is the gross domestic product of the state divided by the population of the state. Now the interesting thing is that Texas, North Dakota, Wyoming, Alaska. and Colorado have oil revenue, which is concentrated in too few hands to provide overall benefit to the state as much as manufacturing would. If you remove the oil revenue, then the GDP per person would drop below the national average and place them solidly into the overall pattern. The states are color-coded with how they voted in the 2020 election with dark gray for those that have consistently voted Republican, white for Democrat, and light gray that switched in 2020 from Republican to Democrat.

The conclusion here is that states that have low GDP per citizen also have the highest levels of gun homicides and suicides.

State or Territory	2020 Deaths per 100,000 people			Poverty Rate		Average GDP	Graduation %	
	Guns	Murder	Suicide	2021	2018	$65,297.52 $GDP/Person	High School	College
Hawaii	3.46	3.2	12.9	10.1%	9.26%	$67,622.14	93.3	35.5
Massachu-setts	3.83	2.6	8.4	7.9%	9.85%	$86,556.81	91.3	46.9
New Jersey	4.78	3.9	7.1	7.4%	9.67%	$71,467.01	90.9	43.1
Rhode Island	4.92	2.6	8.5	9.0%	11.58%	$58,416.16	90.1	38.0
New York	5.30	4.4	8.0	12.3%	13.58%	$91,102.12	87.8	39.5
Connecticut	6.04	4.2	9.3	9.2%	9.78%	$80,729.04	91.6	42.4
California	8.81	6.0	10.0	11.0%	12.58%	$79,286.87	84.4	36.9
Minnesota	8.98	3.5	13.1	7.0%	9.33%	$68,050.01	93.8	37.9
New Hamp-shire	9.23	1.0	16.4	5.6%	7.42%	$64,450.53	94.1	40.2
Nebraska	10.03	3.9	14.9	8.4%	10.37%	$67,210.12	91.9	33.3
Iowa	10.98	3.3	18.0	9.5%	11.11%	$61,696.92	93.1	29.5
Maine	11.11	1.5	16.4	9.2%	11.07%	$50,376.80	93.6	33.5
Washington	11.16	4.2	15.2	7.6%	10.19%	$80,499.69	92.1	38.4
Vermont	11.75	2.2	18.1	8.2%	10.78%	$54,509.61	94.2	42.1
Wisconsin	12.19	5.7	14.5	8.6%	10.97%	$60,012.10	93.1	31.8
Utah	12.85	2.9	20.8	7.5%	9.13%	$60,050.44	94.0	36.9
North Dakota	12.85	3.8	18.2	9.1%	10.53%	$75,034.45	93.5	31.8
Maryland	13.00	10.5	9.2	8.0%	9.02%	$70,587.16	91.1	43.1
South Dakota	13.39	5.9	21.0	10.2%	12.81%	$62,104.04	92.2	28.4
Delaware	13.44	8.7	12.3	9.6%	11.44%	$79,159.22	91.0	34.7
Pennsylvania	13.46	7.6	12.6	10.0%	11.95%	$63,172.80	91.6	34.0
Virginia	13.56	6.1	13.5	8.8%	10.01%	$65,245.62	91.4	42.0
Illinois	13.75	10.6	10.5	9.3%	11.99%	$69,886.01	90.4	37.6
Oregon	13.91	3.7	18.3	9.0%	12.36%	$60,132.53	91.9	36.3
Florida	13.93	7.1	13.2	12.5%	13.34%	$51,518.47	89.6	33.7
Texas	14.09	7.6	13.3	12.9%	14.22%	$63,588.44	85.8	33.2
Michigan	14.49	8.1	14.0	11.0%	13.71%	$53,759.49	92.0	32.1
Ohio	14.99	8.5	13.8	12.3%	13.62%	$59,488.04	91.5	30.6
Colorado	15.87	5.8	21.5	8.9%	9.78%	$68,241.71	92.7	44.2
North Car-olina	16.08	8.5	13.2	12.8%	13.98%	$56,406.96	90.3	34.8

Kansas	16.81	6.6	18.4	8.6%	11.44%	$60,581.56	91.8	35.1
Idaho	16.86	2.3	23.2	8.5%	11.94%	$46,817.27	91.3	30.9
Indiana	17.01	9.1	15.0	10.9%	12.91%	$56,398.06	90.2	28.9
Nevada	17.38	7.0	18.2	12.1%	12.78%	$57,854.02	87.2	28.0
Arizona	17.41	7.3	17.6	11.2%	14.12%	$50,849.50	89.1	33.0
Georgia	17.58	10.2	13.7	13.1%	14.28%	$58,932.72	89.3	34.8
West Virginia	18.20	6.4	19.4	15.0%	17.10%	$44,005.26	89.1	23.1
Kentucky	20.02	9.0	17.7	14.6%	16.61%	$48,212.77	88.7	27.4
Oklahoma	20.70	8.6	21.9	13.8%	15.27%	$51,058.27	89.2	27.0
Tennessee	21.14	10.9	17.2	12.2%	14.62%	$55,143.18	89.7	30.7
Montana	21.51	6.0	26.1	10.4%	12.78%	$49,528.15	94.4	34.6
South Carolina	21.78	12.1	16.3	14.1%	14.68%	$48,078.76	89.4	31.7
Arkansas	22.22	12.3	19.0	15.1%	16.08%	$43,393.84	88.2	24.9
Alabama	22.59	13.0	16.0	14.6%	15.98%	$46,529.47	88.0	27.8
New Mexico	22.63	10.2	24.2	16.7%	18.55%	$50,144.00	87.8	30.1
Missouri	23.11	13.0	18.2	10.8%	13.01%	$53,507.85	91.4	31.9
Alaska	23.84	7.5	27.5	11.7%	10.34%	$74,343.48	93.7	31.9
Louisiana	25.57	18.8	13.7	17.2%	18.65%	$55,265.62	86.9	27.2
Wyoming	26.58	4.3	30.5	9.4%	10.76%	$69,839.26	93.8	28.2
Mississippi	27.73	19.5	13.9	18.1%	19.58%	$38,966.90	86.8	24.5

TABLE 12. GUN DEATHS PER STATE

Further Reading

The following books were consulted before and during the writing of this book. This is not an endorsement of the opinions and conclusions in these books, but an acknowledgement that they were a reference.

Chapter 3: Climate Change

Gates, Bill. *How to avoid a climate disaster: The solutions we have and the breakthroughs we need.* Alfred A Knopf, 2021.

Shellenberger, Michael. *Apocalypse Never: Why environmental alarmism hurts us all.* HarperCollins, 2020.

Chapter 4: Personal Responsibility

Barry, John M. *The Great Influenza: The story of the deadliest pandemic in history*. Penguin Books, 2018.

Chapter 10: Racism

Wilkerson, Isabel. *Caste: The origins of our discontents.* Random House, 2020.

Chapter 17: Fake News

Klein, Ezra. *Why We're Polarized.* Avid Reader Press, 2020.

Miller, Tim. *Why We Did It: A travelogue from the Republican road to hell.* Harper Collins, 2022.

Chapter 22: The Fate of Oil

Yergin, David. *The New Map: Energy, climate, and the clash of nations.* Penguin Press, 2020.

Chapter 27: Healthcare in All Its Forms

Reid, T.R. . *The Healing of America; A global quest for better, cheaper, and fairer healthcare*. Penguin Books, 2010.

Markay, Marty MD. *The Price We Pay: What broke American health care and how to fix it.* Bloomsbury Publishing, 2019.

Rosenthal, Elisabeth MD. *An American Sickness: How healthcare became big business and how you can take it back*. Penguin Books, 2018.

CHAPTER 29: OBESITY AND THE AMERICAN FARM

Pollan, Michael. *The Omnivore's Dilemma: A natural history of four meals.* Penguin Books, 2006.

Salatin, Joel. *Pastured Poultry Profit$*. Polyface, 1996.

Salatin, Joel. *Polyface Micro: Success with Livestock on a Homestead Scale*. Polyface, 2022.

landinstitute.org, The Land Institute

CHAPTER 30: THE DREADED T WORD

Saez, Emmanuel, and Gabriel Zucman. *The Triumph of Injustice: How the Rich Dodge taxes and how to make them pay.* W. W. Norton & Company, 2020.

CHAPTER 31: SUPPLY-SIDE ECONOMICS

Richardson, Heather Cox. *To Make Men Free: A history of the Republican Party.* Basic Books, 2021.

CHAPTER 33: UNIVERSAL BASIC INCOME

Yang, Andrew. *Forward: Notes on the future of our democracy.* Crown Publishing, 2021.

Chapter 46: The Big Problems

Desmond, Matthew. *Poverty by America.* Crown Publishing, 2023.

Chapter 47: The Final Words

Achorn, Edward. *The Lincoln Miracle: Inside the Republican convention that changed history*. Atlantic Monthly Press, 2023

Not Specific to a Chapter

Williams, Joan C. . *White Working Class: Overcoming class cluelessness in America.* Harvard Business Review Press, 2017.

Thaler, Richard H. and Sunstein, Carr R. . *Nudge: The final edition.* Penguin Books, 2021.

Hamilton, Alexander, Madison, James, and Jay, John. *The Federalist Papers; A collection of essays written in favour of the new Constitution.* Oxford University Press, 2008.

Issenberg, Sasha. *The Victory Lab: The secret science of winning campaigns*. Broadway Books, 2016.

King, Bill. *Unapologetically Moderate: My search for a rational center in American politics.* Elite Online Publishing, 2018.

Klaas, Brian. *Corruptible: Who gets power and how it changes us.* Scribner, 2021.

Westen, Drew. *The Political Brain: The role of emotion in deciding the fate of the nation.* PublicAffairs, 2007.

About the Author

William Taggart, a Houstonian by way of New Orleans, has spent more than thirty years as a professional engineer in industry. He has travelled the world, building some of industry's biggest projects and solving some of their thorniest problems. From facilities located in rural America to giant offshore oil platforms in the Gulf of Mexico to other locations scattered across the world, he has seen the diversity of America and the infrastructure that supports civilization.

Now, he has applied analytical skills developed in industry, coupled with his natural inquisitiveness and thirst for creative solutions, to the problems of modern-day American politics. Seeking answers to questions many of us have been asking, he has focused his keen mind on researching these, and other big issues. He invites everyone to think about what America really needs.

Endnotes

1 Gramlich, John. "What the 2020 Electorate Looks like by Party, Race and Ethnicity, Age, Education and Religion." Pew Research Center, October 26, 2020. pewrsr.ch/2TpQBnx.

2 Ball, Molly. "Moderates: Who Are They, and What Do They Want?" Atlantic, May 2014

3 "How much electricity is lost in electricity transmission and distribution in the United States?" EIA.gov. Accessed December 2020. eia.gov/tools/faqs/faq.php?id=105&t=3

4 Smith, Niall. "US tops list of countries fuelling the waste crisis," Verisk Maplecroft, July 2, 2019. maplecroft.com/insights/analysis/us-tops-list-of-countries-fuelling-the-mounting-waste-crisis/

5 "NFIP Debt." FEMA.gov. Accessed August 2, 2023. fema.gov/case-study/nfip-debt.

6 Haskins, Ron. "Three Simple Rules Poor Teens Should Follow to Join the Middle Class," Brookings Institute, March 13, 2013, brookings.edu/opinions/three-simple-rules-poor-teens-should-follow-to-join-the-middle-class/

7 "Federal Debt: A Primer" Congressional Budget Office, March 2020, cbo.gov/system/files/2020-03/56165-CBO-debt-primer.pdf

8 Kortsmit K, Jatlaoui TC, Mandel MG, et al. Abortion Surveillance — United States, 2018. MMWR Surveill Summ 2020;69(No. SS-

7):1–29. DOI: http://dx.doi.org/10.15585/mmwr.ss6907a1external icon.

9 Jones RK, Witwer E, Jerman J. Abortion incidence and service availability in the United States, 2017. New York, NY: Guttmacher Institute; 2019. guttmacher.org/report/abortion-incidence-service-availability-us-2017external icon

10 "Vital Statistics Rapid Release", CDC, Report No. 008, May 2020, cdc.gov/nchs/data/vsrr/vsrr-8-508.pdf

11 "Population and the American Future", Rockefeller Commission Report, 1972, nixonlibrary.gov/finding-aids/fg-275-commission-population-growth-and-american-future-white-house-central-files

12 USA Facts. - usafacts.org/data/topics/people-society/immigration/immigration-and-immigration-enforcement/green-cards-granted/

13 Levitt, Steven and Donohue, John. "The Impact of Legalized Abortion on Crime". The Quarterly Journal of Economics, May 2001.

14 Reed, Ritchie H. and McIntosh , Susan. "Costs of Children" prepared for the Rockefeller Commission Report "Population and the American Future", 1972.

15 "The Cost of Raising a Child." US Department of Agriculture. Accessed May 1, 2020. usda.gov/media/blog/2017/01/13/cost-raising-child

16 Fertility of Women in the United States: 2016, Table 7: Household Income and Income per Household Member Among Women with a Birth in the Past Year, by Marital Status: 2015, US Census Bureau.

census.gov/data/tables/2016/demo/fertility/women-fertility.html#par_reference_1

17 Cohen, Susan A. "Abortion Increasingly Concentrated in World's Poorest Countries.", Guttmacher Institute guttmacher.org/gpr/2012/03/abortion-increasingly-concentrated-worlds-poorest-countries

18 "Public Opinion on Abortion", Pew Research Center, August 29, 2019. pewforum.org/fact-sheet/public-opinion-on-abortion/

19 New York Times Magazine. nytimes.com/interactive/2019/08/14/magazine/1619-america-slavery.html

20 Staff, Washington Post. "Police Shootings Database 2015-2023: Search by Race, Age, Department." The Washington Post, December 5, 2022. washingtonpost.com/graphics/investigations/police-shootings-database/.

21 Thompson, Cheryl W. . "Fatal Police Shootings of Unarmed Black People Reveal Troubling Patterns," NPR, January 25, 2021, npr.org/2021/01/25/956177021/fatal-police-shootings-of-unarmed-black-people-reveal-troubling-patterns.

22 Light, Michael T., He, Jingying, and Robey, Jason P. "Comparing crime rates between undocumented immigrants, legal immigrants, and native-born US citizens in Texas", Proceedings of the National Academy of Sciences Dec 2020, 117 (51) 32340-32347; DOI: 10.1073/pnas.2014704117

23 Brannon, Ike and Albright, Logan. "The Economic and Fiscal Impact of Repealing DACA", Cato Institute, January 18, 2017, cato.org/blog/economic-fiscal-impact-repealing-daca

24 United States v. Wong Kim Ark, 169 US 649 (1898).

25 "Means Matter:Lethality of Suicide Methods," Harvard School of Public Health, January 6, 2017. hsph.harvard.edu/means-matter/means-matter/case-fatality/.

26 Kegler SR, Simon TR, Zwald ML, et al. "Vital Signs: Changes in Firearm Homicide and Suicide Rates — United States, 2019–2020". MMWR Morb Mortal Wkly Rep 2022;71:656–663.

27 Abadi, Mark, Pasley, James, and Ardrey, Taylor. "The 30 deadliest mass shootings in modern US history include Buffalo and Uvalde", Business Insider, May 26, 2022, businessinsider.com/deadliest-mass-shootings-in-us-history-2017-10

28 Campbell, Richard. Structure Fires in Schools, National Fire Protection Association. September 2020, nfpa.org/News-and-Research/Data-research-and-tools/Building-and-Life-Safety/Structure-fires-in-schools

29 Poushter, Jacob and Kent, Nicholas. "The Global Divide on Homosexuality Persists", Pew Research Center. June 25, 2020, pewresearch.org/global/2020/06/25/global-divide-on-homosexuality-persists/

30 MASTERPIECE CAKESHOP, LTD., ET AL. v. COLORADO CIVIL RIGHTS COMMISSION ET AL., supremecourt.gov/opinions/17pdf/16-111_j4el.pdf

31 303 CREATIVE LLC ET AL. v. ELENIS ET AL. CERTIORARI TO THE UNITED STATES COURT OF APPEALS FOR THE TENTH CIRCUIT No. 21–476. Argued December 5, 2022—Decided June 30, 2023; supremecourt.gov/opinions/22pdf/21-476_c185.pdf

32 Bustos, Valeria P. MD; Bustos, Samyd S. MD; Mascaro, Andres MD; Del Corral, Gabriel MD, FACS; Forte, Antonio J. MD, PhD, MS; Ciudad, Pedro MD, PhD; Kim, Esther A. MD; Langstein, Howard N. MD; Manrique, Oscar J. MD, FACS, "Regret after Gender-affirmation Surgery: A Systematic Review and Meta-analysis of Prevalence, Plastic and Reconstructive Surgery" - Global Open: March 2021 - Volume 9 - Issue 3 - p e3477 doi: 10.1097/GOX.0000000000003477

33 Maldonado, Camilo. "Price of College Increasing Almost 8 Times Faster than Wages." Forbes. *Forbes Magazine*, June 24, 2018. forbes.com/sites/camilomaldonado/2018/07/24/price-of-college-increasing-almost-8-times-faster-than-wages/?sh=4094a00f66c1.

34 Miller, Ben, Campbell, Colleen, Cohen, Brent J., and Hancock, Charlotte. "Addressing the $1.5 Trillion in Federal Student Loan Debt", Center for American Progress June 12, 2019, americanprogress.org/issues/education-postsecondary/reports/2019/06/12/470893/addressing-1-5-trillion-federal-student-loan-debt/

35 Jimenez, L., Sargrad, S., Morales, J., and Thompson, M. . "Remedial Education, The cost of catching up", Center for American Progress, September 2016, cdn.americanprogress.org/content/uploads/2016/09/29120402/CostOfCatchingUp2-report.pdf?_ga=2.32006654.1121059578.1610034651-637292997.1608240225

36 Desilver, Drew. "US students' academic achievement still lags that of their peers in many other countries", Pew Research Center, February 15, 2017, pewresearch.org/short-reads/2017/02/15/u-s-students-internationally-math-science/

37 "Private schools show same results as public schools", University of Queensland, April 2015, uq.edu.au/news/article/2015/04/private-schools-show-same-results-public-schools

38 Mongeau, Lillian. "The Never-Ending Struggle to Improve Head Start", The Atlantic, August 9, 2016, theatlantic.com/education/archive/2016/08/is-head-start-a-failure/494942/

39 "America After 3PM Special Report: Afterschool Programs in Demand," After School Alliance, October 2014, afterschoolalliance.org/documents/AA3PM-2014/AA3PM_Key_Findings.pdf

40 "All In Together: Creating Places Where Young And Old Thrive", Eisner Foundation, eisnerfoundation.org/resources/additional-resources/

41 Liptak, Adam. "The Ads Discriminate, but Does the Web?" *New York Times*, March 8, 2006

42 Letter from the National Association of Attorneys General to Congress, dated July 23, 2013, eff.org/sites/default/files/cda-ag-letter.pdf

43 Rather, Dan, January 22, 2017, "Dan Rather Facebook post". Facebook.

44 Palazzolo, Joe. "We Won't See You in Court: The Era of Tort Lawsuits Is Waning," *Wall Street Journal*, July 24, 2017, wsj.com/arti-

cles/we-wont-see-you-in-court-the-era-of-tort-lawsuits-is-waning-1500930572

45 Glynn, S. ."TOXIC TOYS AND DANGEROUS DRYWALL: HOLDING FOREIGN MANUFACTURERS LIABLE FOR DEFECTIVE PRODUCTS—THE FUND CONCEPT," *Emory Law Review,* Vol. 26, page 317-364, 2012.

46 Sixel, L.M. ."Private Equity Funds Deploy New Investment Strategy: Financing lawsuits,", Texans for Lawsuit Reform, June 6, 2018, tortreform.com/news/private-equity-funds-deploy-new-investment-strategy-financing-lawsuits/.

47 Employment of Coal Mining Industry, US Bureau of Labor Statistics, Series ID CES1021210001, accessed December 2020, data.bls.gov/pdq/SurveyOutputServlet

48 "US energy facts explained", US Energy Information Administration, Accessed December 2020. eia.gov/energyexplained/us-energy-facts/

49 "Capital Cost and Performance Characteristic Estimates for Utility Scale Electric Power Generating Technologies", US Energy Information Administration, February 2020 eia.gov/analysis/studies/powerplants/capitalcost/pdf/capital_cost_AEO2020.pdf

50 "Direct Air Capture", IEA, 2022, Paris iea.org/reports/direct-air-capture

51 "America's Electricity Generation Capacity 2020 Update," American Public Power Association, publicpower.org/system/files/documents/Americas-Electricity-Generation-Capacity-2020.pdf

52 Hydropower Vision: New Report Highlights Future Pathways for US Hydropower, Department of Energy, July 26, 2016, energy.gov/articles/hydropower-vision-new-report-highlights-future-pathways-us-hydropower

53 "Scale-up of Solar and Wind Puts Existing Coal, Gas at Risk". BloombergNEF, 28 April 2020. Retrieved 31 May 2020., about.bnef.com/blog/scale-up-of-solar-and-wind-puts-existing-coal-gas-at-risk/

54 Kutscher, Charles F. . "The Status and Future of Geothermal Electric Power", NREL/CP-550-28204, Presented at the American Solar Energy Society (ASES) Conference Madison, Wisconsin June 16-21, 2000

55 McKenna, J., Blackwell, D., Moyes, C., and Patterson, P.D. . "Geothermal electric power supply possible from Gulf Coast Midcontient oil field waters", Oil & Gas Journal, Sept 5, 2005

56 "US Vehicle Miles", US Department of Transportation, Bureau of Transportation Statistics, bts.gov/content/us-vehicle-miles

57 "Energy Storage Grand Challenge", Department of Energy, energy.gov/energy-storage-grand-challenge/energy-storage-grand-challenge

58 Eaton, C. and Morenne, B. "US Shale Boom Shows Signs of Peaking as Big Oil Wells Disappear," *Wall Street Journal*, March 8, 2023 wsj.com/articles/u-s-shale-boom-shows-signs-of-peaking-as-big-oil-wells-disappear

59 "A major(s) rating! Total outperformed peers in 2020, a year of cost reductions and poor discoveries", Rystad Energy, January 28, 2021, rystadenergy.com/newsevents/news/press-releases/a-majors-rating-

total-outperformed-peers-in-2020-a-year-of-cost-reductions-and-poor-discoveries/

60 Prices compiled through multiple websites - en.wikipedia.org/wiki/Gasoline_and_diesel_usage_and_pricing, tradingeconomics.com/country-list/gasoline-prices, globalpetrolprices.com/China/gasoline_prices/

61 "THE 2022 ANNUAL REPORT OF THE BOARD OF TRUSTEES OF THE FEDERAL OLD-AGE AND SURVIVORS INSURANCE AND FEDERAL DISABILITY INSURANCE TRUST FUNDS", Social Security Administration, released June 2022, ssa.gov/OACT/tr/2022/tr2022.pdf

62 "Social Security," Individual Changes Modifying Social Security, accessed July 15, 2023, ssa.gov/OACT/solvency/provisions/.

63 Duncan, Ian et al. "Medicare Cost at End of Life." The American journal of hospice & palliative care vol. 36,8 (2019): 705-710. doi:10.1177/1049909119836204

64 Allen, Kent, "Taking a Second Look at End-of-Life Medicare Health Costs", American Association of Retired Persons, July 5, 2018, aarp.org/health/medicare-insurance/info-2018/medicare-spending-on-dying-patients.html

65 Aldridge MD, Kelley AS. The Myth Regarding the High Cost of End-of-Life Care. Am J Public Health. 2015;105(12):2411-2415. doi:10.2105/AJPH.2015.302889

66 Peterson-KFF Health System Tracker, healthsystemtracker.org/chart-collection/health-spending-u-s-compare-coun-

tries/#item-spendingcomparison_health-consumption-expenditures-per-capita-2019

67 List of countries by life expectancy, en.wikipedia.org/wiki/List_of_countries_by_life_expectancy

68 Himmelstein, David U., Lawless, Robert M., Thorne, Deborah, Foohey, Pamela, and Woolhandler, Steffie. "2019: Medical Bankruptcy: Still Common Despite the Affordable Care Act", American Journal of Public Health 109, 431_433, doi.org/10.2105/AJPH.2018.304901

69 Markay, Marty MD. The Price We Pay: What broke american health care and how to fix it. Bloomsbury Publishing, 2019, page 29.

70 Markay, Marty MD. The Price We Pay: What broke american health care and how to fix it. Bloomsbury Publishing, 2019, page 30.

71 Wates, H and Graf, M. . "The Cost of Chronic Disease in the US", Milken Institute, August 28, 2018, milkeninstitute.org/report/costs-chronic-disease-us

72 "Adult Obesity Facts", Center for Disease Control and Prevention, September 30, 2021, cdc.gov/obesity/data/adult.html

73 McGinty, Tom, Mathews, Anna Wilde and Evans, Melanie."Hospitals Hide Pricing Data from Search Results", Wall Street Journal, March 22, 2021, wsj.com/articles/hospitals-hide-pricing-data-from-search-results-11616405402

74 "Public Opinion on Single-Payer, National Health Plans, and Expanding Access to Medicare Coverage", Kaiser Family Foundation, Oct 16, 2020. kff.org/slideshow/public-opinion-on-single-payer-national-health-plans-and-expanding-access-to-medicare-coverage/

"BIDEN RETAKES LEAD AS WARREN PLUNGES, BUTTIGIEG RISES, QUINNIPIAC UNIVERSITY NATIONAL POLL FINDS; VOTERS NOT SWAYED BY IMPEACHMENT HEARINGS". poll.qu.edu. November 26, 2019. Retrieved January 22, 2019. poll.qu.edu/Poll-Release-Legacy?releaseid=3650

75 "Increase Excise Taxes on Tobacco Products", Congressional Budget Office, cbo.gov/budget-options/56869

76 "The Astronomical Price of Insulin Hurts American Families", Rand Corporation, January 6th, 2021, rand.org/blog/rand-review/2021/01/the-astronomical-price-of-insulin-hurts-american-families.html

77 Gotham D, Barber MJ, Hill. "A Production costs and potential prices for biosimilars of human insulin and insulin analogues", BMJ Global Health 2018;3:e000850.

78 Rosenthal, Elisabeth MD. *An American Sickness: How healthcare became big business and how you can take it back.* Penguin Books, 2018. Pages 106-107

79 Markay, Marty MD. *The Price We Pay: What broke American health care and how to fix it.* Bloomsbury Publishing, 2019, pages 191-204.

80 Safer DJ. "Overprescribed Medications for US Adults: Four Major Examples.", J Clin Med Res. 2019 Sep;11(9):617-622. doi: 10.14740/jocmr3906. Epub 2019 Sep 1. PMID: 31523334; PMCID: PMC6731049.

81 Sheikh-Taha, M., Asmar, M., "Polypharmacy and severe potential drug-drug interactions among older adults with cardiovascular disease

in the United States.", BMC Geriatr 21, 233 (2021). doi.org/10.1186/s12877-021-02183-0

82 Xie L, Gelfand A, Murphy CC, Mathew MS, Atem F, Delclos GL, Messiah S. "Prevalence of polypharmacy and associated adverse outcomes and risk factors among children with asthma in the USA: a cross-sectional study." BMJ Open. 2022 Oct 13;12(10):e064708. doi: 10.1136/bmjopen-2022-064708. PMID: 36229143; PMCID: PMC9562747.

83 "World population with and without synthetic nitrogen fertilizers", ourworldindata.org/grapher/world-population-with-and-without-fertilizer?country=~OWID_WRL

84 Pieper, M., Michalke, A. & Gaugler, T., "Calculation of external climate costs for food highlights inadequate pricing of animal products.", Nat Commun 11, 6117 (2020). doi.org/10.1038/s41467-020-19474-6

85 "Grain-Fed Beef vs. Grass-Fed Beef – Greenhouse Gas Emissions", http://newzealmeats.com/blog/grain-fed-vs-grass-fed-beef-greenhouse-gas-emissions/

86 For Federal Tax Brackets through history - tax-brackets.org/federaltaxtable/ For calculating inflation - in2013dollars.com/us/inflation/

87 Data on Net Worth from the Federal Reserve of St. Louis, fred.stlouisfed.org/categories/33001 analyzed by the author.

88 "Most Americans Say There Is Too Much Economic Inequality in the US, but Fewer Than Half Call It a Top Priority", Pew Research Center, January 2020

89 The Distribution of Household Income, Congressional Budget Office, 2019, cbo.gov/system/files/2022-11/58353-HouseholdIncome.pdf

90 Kiel, Paul and Eisinger, Jesse. "How the IRS Was Gutted", ProPublica, Dec. 11, 2018, propublica.org/article/how-the-irs-was-gutted

91 Richardson, Heather Cox. *To Make Men Free: A history of the Republican party*. Basic Books, 2021, Page 173.

92 Simpson, Bob. "How Sending Stimulus Checks to the Poor Can Boost the US Economy", Chicago Booth Review, April 4, 2022, chicagobooth.edu/review/how-sending-stimulus-checks-poor-can-boost-us-economyReferencing research - Joel P. Flynn, Christina Patterson, and John Sturm, "Fiscal Policy in a Networked Economy," Working paper, January 2022.

93 Cost of Living Data Series, meric.mo.gov/data/cost-living-data-series

94 Prang, Allison. "$15-an-Hour Minimum Wage Could Further Sting Teen Employment", Wall Street Journal, March 27, 2021, wsj.com/articles/15-an-hour-minimum-wage-could-further-sting-teen-employment-11616837401

95 US Bureau of Labor Statistics, bls.gov/opub/reports/minimum-wage/2019/pdf/home.pdf

96 "The Budgetary Effects of the Raise the Wage Act of 2021", Congressional Budget Office, February 8, 2021, cbo.gov/publication/56975

97 "Despite A Rough Two Years, US Independent Restaurants Still Represent Over Half of US Commercial Restaurant Units and Are Recovering", NPD Group, May 12, 2022, npd.com/news/press-releases/2022/despite-a-rough-two-years-u-s-independent-restaurants-still-represent-over-half-of-u-s-commercial-restaurant-units-and-are-recovering/

98 Wells, K., Attoh, K., and Cullen, D. . "Working Class Perspective: The Work Lives of Uber Drivers Are Worse Than You Think", Georgetown University, July 10, 2017, lwp.georgetown.edu/visitingscholars/wcp-the-work-lives-of-uber-drivers-is-worse-than-you-think/

99 United States Census – Table S1901 from 2019 American Community Survey 1-Year Estimates, data.census.gov/cedsci/table?q=United%20States&g=0100000US&tid=ACSST1Y2019.S1901&hidePreview=true

100 Jones, Damon and Marinescu, Ioana Elena, "The Labor Market Impacts of Universal and Permanent Cash Transfers: Evidence from the Alaska Permanent Fund (February 5, 2018)", Available at SSRN: ssrn.com/abstract=3118343 or http://dx.doi.org/10.2139/ssrn.3118343

101 "Financial Accounts of the United States", Federal Reserve, federalreserve.gov/releases/z1/20210923/html/introductory_text.htm

102 "Industrial Tariffs", office of United States Trade Representative, ustr.gov/issue-areas/industry-manufacturing/industrial-tariffs

103 "The People's Republic of China", Office of the United States Trade Representative, ustr.gov/countries-regions/china-mongolia-taiwan/peoples-republic-china

104 Curiel, John A. and Steelman, Tyler. "Redistricting out Representation: Democratic Harms in Splitting Zip Codes", July 30, 2018, http://jacuriel.web.unc.edu/files/2018/08/redist_out_repsv7.pdf

105 "How Accurate is Matching ZIP Codes to Legislative Districts?" azavea.com/blog/2017/07/26/accuracy-zip-district-matching/

106 Massoglia, Anna. "Cases show foreign donors secretly funnel money through straw donors, shell companies, 'dark money'", Open Secrets, October 28, 2021, opensecrets.org/news/2021/10/foreign-donors-funnel-money-straw-donors-shell-companies-dark-money/

107 Vogel, Kenneth and Goldmacher, Shane. "Democrats Decried Dark Money. Then They Won With It in 2020", New York Times, updated Aug 21, 2022, nytimes.com/2022/01/29/us/politics/democrats-dark-money-donors.html

108 Richardson, Heather Cox. To Make Men Free: A history of the Republican party. Basic Books, 2021, Page 454.

109 Salvanto, Anthony. "Americans continue to feel US democracy is under threat — CBS News poll", CBS News, September 1, 2022, cbsnews.com/news/cbs-news-poll-americans-democracy-is-under-threat-opinion-poll-2022-09-01/

110 Casey JA, Su JG, Henneman LRF, Zigler C, Neophytou AM, Catalano R, Gondalia R, Chen YT, Kaye L, Moyer SS, Combs V, Simrall G, Smith T, Sublett J, Barrett MA., "Improved asthma outcomes observed in the vicinity of coal power plant retirement, retrofit, and conversion to natural gas.", Nat Energy. 2020 May;5(5):398-408. doi: 10.1038/s41560-020-0600-2. Epub 2020 Apr 13. PMID: 32483491; PMCID: PMC7263319.

111 Flax, Bill. "The True Meaning of Separation of Church and State", Forbes, July 9, 2011, forbes.com/sites/billflax/2011/07/09/the-true-meaning-of-separation-of-church-and-state/

112 Bivens, Josh and Kandra, Jori. "CEO pay has skyrocketed 1,460% since 1978," Economic Policy Institute, October 2022, epi.org/publication/ceo-pay-in-2021/

113 Firearm Mortality by State, Center for Disease Control and Prevention, cdc.gov/nchs/pressroom/sosmap/firearm_mortality/firearm.htm